Topics in Biodiversity and Conservation

Volume 17

More information about this series at http://www.springer.com/series/7488

F.P.G. Princée

Exploring Studbooks for Wildlife Management and Conservation

Springer

F.P.G. Princée
Population Management and Wildlife Conservation
East Rudham, Norfolk, UK

ISSN 1875-1288 ISSN 1875-1296 (electronic)
Topics in Biodiversity and Conservation
ISBN 978-3-319-50031-7 ISBN 978-3-319-50032-4 (eBook)
DOI 10.1007/978-3-319-50032-4

Library of Congress Control Number: 2016961290

Printed on acid-free paper

This Springer imprint is published by Springer Nature
The registered company is Springer International Publishing AG
The registered company address is: Gewerbestrasse 11, 6330 Cham, Switzerland

Preface

Some 35 years ago, I became involved in conservation of endangered animal species. It all started while I was a biology student at the Biological Department of the Royal Rotterdam Zoological and Botanical Gardens ("Blijdorp Zoo"). It was around the time that the international zoo world introduced cooperative species management programmes, based on demographic and genetic science, in order to establish self-sustaining populations of endangered species. This created an environment in which applied and fundamental research were intertwined and which shaped my scientific background in population genetics.

The introduction of personal computers opened a "new" creative platform that allowed me to develop the skills of computer programming. This resulted in developing simulation models to explore population genetic processes in real populations with their complex pedigrees. The first computer program that I developed was followed by the next, and the next, etc. resulting in studbook software that included a range of demographic and genetic analyses.

The catalyst for writing this book dates back to the years that I worked for the Dutch National Foundation for Research in Zoological Gardens/European Association of Zoos and Aquaria. Teaching, training and advising coordinators of European Endangered Species Programmes (EEPs) in population biology and studbook software meant exposure to a diversity of practical problems in species management. Solutions to some of these problems may be embedded in the studbook itself, and that is where exploring starts.

I participated in various workshops of the IUCN/SSC/Conservation Breeding Specialist Group in which the knowledge of zoo and wildlife professionals was exchanged and combined. These workshops, the Population and Habitat Viability Assessment of the Aruba island rattlesnake in particular, exposed me to the reality of field conservation.

I became more directly involved in nature conservation projects in the 1990s through the conservation fund of the Dutch zoos. The induction was a population viability study on lions in the Queen Elizabeth National Park (Uganda) that was carried out by Makerere University (Kampala). Individual identification of lions, a small population and urbanised areas acting as a barrier to movement do not make it difficult to foresee studbook-style management in the future.

A few years later, I started to write the first chapters while living next to Ranomafana National Park, a rainforest in the south-east of Madagascar. No zoo environment, but wild nature. However, the state of some endangered species in this 42,000 ha protected area highlighted that nature is (gradually) becoming a big zoo. Take, for example, the closely monitored greater bamboo lemur population in this park at that time. One day there were six animals, the next day one animal was "missing", and one month later a young was born, bringing the group total back to six. This doesn't sound like a wild population but more closely resembles a zoo group. Field conservation is exploring too, as researchers discovered four new groups, totalling more than 100 individuals, in fragmented bamboo forests within the peripheral zone of the park in 2009 and 2010.

The idea of exploring studbook data for its relevance to conservation in the natural habitat was reinforced while I was living in Dzanga Sangha Protected Areas (Central African Republic). The secretive lifestyle of many of the inhabitants of this tropical rainforest not only makes it difficult to assess population sizes but also to obtain natural history data that are required for conservation measures. My background prompted me to ask zoo colleagues for studbook data in order to extract basic biological information on species in the protected area. I have included a PVA study on lowland bongo as an example in this book.

Exploring studbook data "forced" me to explore statistics in more depth than I had done before. Studbooks are relatively small and the issue of reliability of estimated parameters cannot be ignored. Moreover, problem-solving involves comparative studies that simply require significance tests. However, this is not a book on studbook statistics. The main focus is exploring the wealth of (biological) data in studbooks, in which statistics can play a supporting role.

I have selected topics in demographics and genetics of studbook populations that deserve more attention, but which can be studied in standard data sets. These topics are interconnected, which means that, for example, life tables are used in tests for inbreeding depression and litter size is considered as a "trait" in quantitative genetics.

This book is intended for professionals in zoo and wildlife conservation who are involved in management of captive populations; reintroduction or management of small, isolated, fragmented populations in the wild; and for researchers and students who are planning studies that are based on studbook populations or who want to explore what studbook data can offer for topics in academic research.

East Rudham, UK F.P.G. Princée
September, 2016

Acknowledgements

A monograph is the work of a single author but not of a single mind. This book is the result of experiences and ideas that have been accumulated over years and have been shaped by discussions with colleagues, from zoo keeper to director and from ranger to game warden, as well as with colleagues in governments, NGOs and universities.

First of all, I would like to thank Anna Feistner for reading this book through several times with an editorial eye and for her comments and ideas as a conservation biologist. I am indebted to Anna for her encouragement, support and trust in enabling me to realise a book which took considerably more time than I had once optimistically foreseen. I thank Bert de Boer for his valuable comments on an earlier draft and ideas that helped in streamlining this book.

I could not have written this book without access to real-time computerised studbooks on endangered species. I thank the following species managers/studbook keepers and the institutions to whom they are or were affiliated: Richard Barnes (Port Lympne Reserve, UK), Rob Belterman (Rotterdam Zoo, The Netherlands), Leif Blomqvist (Nordic Ark, Sweden, and Helsinki Zoo, Finland), Lydia Frazier Bosley (Logsden, USA), Angela Glatston (Rotterdam Zoo, The Netherlands), Gerard Meijer (Ouwehands Zoo, The Netherlands), Hanny Verberkmoes and Wim Verberkmoes (GaiaZOO, The Netherlands) for providing their studbook data to be used as examples in this book. I acknowledge Andrea Turkalo (the Wildlife Conservation Society, USA) for field data on lowland bongo in Dzanga Baie.

I would like to thank Jonathan Ballou, Laurie Bingaman-Lackey, Lisa Faust, Robert Lacy, Nathan Flesness and Andy Odum for discussions, some of which started years ago, on various topics in demography, genetics and studbook software. Erik Postma is acknowledged for explaining assumptions that are made regarding missing parents in the animal model.

Ludwig Siefert is acknowledged for introducing me to the reality of field conservation when I was involved in supporting the Ugandan lion/carnivore project. I thank Jean Claude Razafimahaimodison for all discussions when I exchanged the comfort of analysing zoo populations for interpreting transect data from Ranomafana National Park. Angelique Todd and Andrea Turkalo are acknowledged for introducing me to the world of gorillas, forest elephants and lowland bongo.

Contents

Part II Demographic Analyses

Symbols

Natural History

Ω	Longevity, i.e. oldest age observed
T_1	First age at reproduction

Census

N_t	Population size at census t
B_t	Number of births in interval t
D_t	Number of deaths in interval t
E_t	Number of emigrants in interval t
I_t	Number of immigrants in interval t
b_t	Crude birth rate in interval t
d_t	Crude death rate in interval t
e_t	Crude emigration rate in interval t
i_t	Crude immigration rate in interval t

Life Tables

n_x	Number of individuals in age class x
d_x	Crude rate, number of individuals dying during the age interval x to $x + 1$
q_x	Mortality rate in age class x
p_x	Survival rate in age class x
l_x	Survivorship or survival rate to start of age class x
L_x	Midpoint survivorship; individuals alive on average during the age interval x to $x + 1$.

e_x	Mean expectation of life for individuals alive at start of age x
m_x	Fecundity rate of individuals in age class x
V_x	Reproductive value of individuals of age x; expected number of future offspring
C_x	Proportion of individuals of age x in the stable age distribution
$\bar{T}$	Population generation time
λ	Finite rate of increase
r	Intrinsic rate of increase
R_0	Net reproductive rate; growth rate per generation time
ω	Oldest age class
σ_i	Probability of surviving in stage i during a time step
γ_i	Probability of moving to the next stage $i+1$ during a time step
G_i	Probability of surviving and moving to next stage $i+1$ during a time step; $G_i = \sigma_i \gamma_i$
P_i	Probability of surviving and remaining in stage i during a time step; $P_i = \sigma_i(1-\gamma_i)$

Genetics

δ	Measure of inbreeding depression
μ	Mutation rate
ϕ	Kinship coefficient
B	Lethal equivalents
F	Fixation index
f	Inbreeding coefficient
H_e	Expected heterozygosity or gene diversity in population
h_e	Expected heterozygosity at a locus
H_o	Observed heterozygosity in population
N_e	Effective population size
N_f	Number of breeding females in population
N_m	Number of breeding males in population
N_{ev}	Variance effective population size
$\bar{n_a}$	Average number of alleles on r loci
P	Proportion of polymorphic loci
r	Number of loci

Quantitative Genetics

CV_P	Phenotypic coefficient of variation
EBV_i	Estimated breeding value of individual i
H^2	Broad-sense heritability

h^2	Narrow-sense heritability
h^2_{adj}	Narrow-sense heritability adjusted for phenotypic correlation
r_p	Repeatability
r_{pp}	Phenotypic correlation between parents (mates)
V_A	Additive genetic variance
V_D	Dominance variance
V_E	Environmental variance
V_{Eg}	General or permanent environmental variance
V_{Es}	Specialised environmental variance
V_I	Epistatic variance
V_P	Phenotypic variance
V_R	Residual variance

Part I
Introduction

Chapter 1
Introduction

Abstract This chapter provides an introduction to this book's goal of exploring studbooks. The development of studbooks is reviewed, starting with the first studbooks for domestic species in the eighteenth and nineteenth centuries, which became data sources for the early studies on inheritance. The history of studbooks on wild species started around the end of the nineteenth century and expanded considerably in the 1960s. The overview continues with the founding of regional zoo management programmes in the 1980s and the transition of studbooks from registers to electronic databases with demographic and genetic analyses. The concept of the "population management triangle" which reflects the interaction between husbandry, demographics and genetics is presented. The section on exploring studbooks, the title of this book, reviews the wealth of data on endangered animal species that is stored in studbooks and that can be used for wildlife management and conservation. An outline of the main topics of the book is provided. These include natural history, census, life tables, survival analysis, population projections, detection of inbreeding depression, quantitative genetics and the use of data from captivity in conservation.

1.1 A Brief History

A *studbook* or *stud book* is described as "a written record of the pedigree of a purebred stock, esp. of racehorses" (McLeod 1987). This reference to racehorses is not a coincidence, as the first studbook known to be published (in 1791) contained pedigrees of Thoroughbred horses that could be traced back to the founders (Weatherby 1791). Some 30 years later, the first studbook or *herd–book* for (short–horn) cattle was published (Coates 1822).

Following the publication of Darwin's *Origin of Species*, scientific methods that used pedigree data were applied to improve livestock, especially through selection within breeds (Brewer 1893). Studbooks also became the data source for genetic studies, e.g. reproductive selection in horses (Pearson et al. 1899); and studies on Mendelian inheritance, e.g. coat colour in horses and dogs (Hurst 1906; Little 1914). The studbooks for domestic species can be considered as the foundation on which studbooks on endangered species, which are the subjects of this book, were established.

F.P.G. Princée, *Exploring Studbooks for Wildlife Management and Conservation*,
Topics in Biodiversity and Conservation 17, DOI 10.1007/978-3-319-50032-4_1

The first studbooks for wild species were for the European bison or wisent (*Bison bonasus*), Père David's deer (*Elaphurus davidianus*) and Przewalski's horse (*Equus przewalskii*) (Mohr 1971; Tong 1957; von der Groeben 1932). The purpose of these studbooks differed from those of domestic species: preservation of species that were on the brink of extinction or already extinct in the wild.

By the early 1920s, the wild population of European bison was in a deplorable state due to the collapse of protection during the First World War and the Russian Revolution (Ahrens 1921). A preservation programme similar to that of the American Bison Society, was proposed (Ahrens 1921, 1923). This resulted in establishing the *Internationale Gesellschaft zur Erhaltung des Wisents* (*International Society for the Protection of the European Bison*). A card catalogue with information on all (then) living wisents had been compiled (Ahrens 1923) and formed the basis for the first edition of the studbook for wisent in 1932 (von der Groeben 1932).

Some 70 years after the Przewalski's horse was described by science (in 1881), its extinction in the wild was feared (Bouman and Bouman 1994). The captive population became critically important as the only way to preserve the species, resulting in the compilation of the first edition of the studbook in 1959 (Mohr 1971; Volf 1994).

In response to the increasing number of endangered animal species, the *International Zoo Yearbook* (Zoological Society of London) initiated and published the first world–wide zoo census of rare wild animals in 1962 (Anonymous 1962). This census included species which were described as "endangered" by the *International Union for Conservation of Nature (IUCN)*. Five years later, the *International Union of Directors of Zoological Gardens (IUDZG)*[1] and the IUCN recommended starting studbooks for rare species of animals in captivity in order to facilitate planned breeding and to establish self–sustaining populations, i.e. populations without import of wild–caught individuals (Anonymous 1968).

1.2 From Register to Management

The *Convention on International Trade in Endangered Species of Wild Fauna and Flora (CITES)* entered into force in July 1975 (Wijnstekers 2011), by which time some 37 international studbooks for rare animals had been established (Anonymous 1975). However, the zoos' goal of establishing self–sustaining populations turned out to be hard to achieve.

A study based on 15 years of census data in the International Zoo Yearbook, showed that only 10 % of the listed species could be considered self–sustaining by the end of the 1970s (Pinder and Barkham 1978). In addition, studies on zoo populations showed the negative impact that inbreeding can have on fecundity and (early) survival (e.g. Flesness 1977; Ralls et al. 1979). The IUDZG introduced new

[1]Renamed to *World Association of Zoos and Aquariums (WAZA)* in 2000.

management guidelines in 1980, that involved a pro-active approach in conducting demographic and genetic analyses on studbooks (Anonymous 1980).

Conservation genetics was "introduced" to the zoo world in the early 1980s (e.g. Frankel and Soulé 1981; Schonewald-Cox et al. 1983). It was recommended to maintain at least 90 % of the original (wild) genetic variation in captive populations over a period of 200 years (Soulé et al. 1986). Simply breeding in numbers was no longer considered sufficient to achieve self–sustaining populations. Minimising genetic loss due to inbreeding and genetic drift required an interrelated approach of both demographic and genetic management (Foose 1983).

Intensive population management includes recommendations such as which individuals should be paired, and, depending on population size, which pairs (or groups) are allowed to breed in the next season(s), and which should not breed. Implementation of this type of management requires close cooperation between individual institutions. Practical reasons (logistics, language, legislation) resulted in zoos establishing regional cooperative management programmes in the 1980s, e.g. the *Species Survival Plan (SSP)* in North America, *Australasian Species Management Programs (ASMP)* of Australasia and the *European Endangered Species Programme (EEP)*[2] in continental Europe (e.g. Lees and Wilcken 2009). By 2011 the number of international and regional studbooks had extended to 190 and 1,350, respectively (Anonymous 2015).

However, even regional zoo populations can be too small and/or based on too few founders to be self–sustaining long term. This requires management at a global level (Lees and Wilcken 2009). A global management plan for the red panda (*Ailurus fulgens*) in which the regional programmes cooperated closely was already developed in 1992 (Glatston and Princée 1993; Princée and Glatston 1992). The need to move from regional to global management when required, was recognised by the *World Association of Zoos and Aquaria (WAZA)* by introducing the *Global Species Management Plan (GMSP)* (WAZA 2016).

Three international and five regional studbooks are used as case examples throughout this book. Chapter 2 provides background information on the species involved.

1.3 Population Management Triangle

Population management programmes may initially have been perceived as being dominated by demographics and genetics. However, any recommendation, whether referring to increasing reproductive success, reducing juvenile mortality or exchanging animals between collections, directly or indirectly points to husbandry, i.e. the science of breeding, rearing, and caring for animals.

[2]EEP is the abbreviation of the original name *Europäisches Erhaltungszuchtprogramme*.

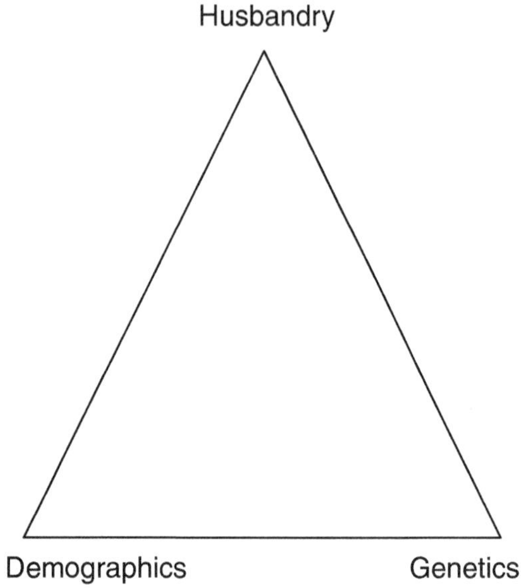

Fig. 1.1 Population management triangle

The interrelationship between husbandry, demographics and genetics is represented by the *Population Management Triangle* in Fig. 1.1. A successful population management programme depends on good husbandry practice with high standards of welfare, allowing the breeding of individuals on a regular and planned basis, following demographic and genetic goals, from generation to generation. Husbandry can affect the success of a population management programme in various ways. For example, reproductive success may be limited to a few institutions, resulting, sooner or later, in inbreeding. This scenario is not uncommon in populations that were established before the introduction of studbooks and population management programmes.

Risks of inbreeding or aggressive behaviour exist when offspring are kept with their parents beyond the age of dispersal or sexual maturity. This is generally not a deliberate decision, but can be due to a variety of factors, including delayed transport permits to another institution, or individuals reaching sexual maturity at an earlier age than previously known.

Demographic analyses can be used in comparative studies to evaluate husbandry practices or to assess vulnerability to inbreeding depression. Results of such studies can lead to improvements in husbandry or in the development of exchanges between regional programmes to keep inbreeding within regions at acceptable (low) levels.

The Population Management Triangle reflects an holistic approach to animal management and is present in the background throughout this book.

1.4 Exploring Studbooks

Computerisation of studbooks in the mid 1980s initiated the development of demographic and genetic analyses programs within the zoo community (see overview in Ballou et al. 2010). Since then, some 30 years of experience has been gained in population management involving a wide range of wild species. The science behind small population management is tested and validated (Lees and Wilcken 2009), with management guidelines produced and regularly updated (e.g. Ballou and Lacy 1995; Ballou et al. 2010; Foose and Ballou 1988). This has resulted in development of population management software in which previously stand–alone analyses are combined e.g. PMx (Ballou et al. 2011).

Species managers and studbook keepers primarily use these software tools to assess the status of a population, to evaluate effects of management measures to achieve the goal of self–sustaining populations, and to make recommendations. However, some studbooks have been kept for 30+ years and thus contain sufficient numbers of records (i.e. sample sizes) to allow for comparative studies that can be tested statistically.

Studbook data can be used to study differences between subgroups within (regional) populations or to study effects of a factor on a given biological parameter, e.g. influence of climate on juvenile survival in red pandas (Princée and Glatston 2016). Studbook analyses do not necessarily need to be aimed at improving captive management. Conservationists are often confronted with limited available knowledge on life history of endangered species. Analyses of studbook data can provide baseline information on species' life history that is essential in *Population and Habitat Viability Assessment (PHVA)*. Studbook data, whether on a single or on multiple species, as in the database of the *International Species Information System (ISIS)*,[3] can be important more generally, e.g. for studies on life expectancy and ageing (Ricklefs 2010; Young et al. 2012).

This book focusses on exploring studbook data that are or could be useful for management, conservation and science in general. An overview is provided in the next section. Chapter 2 presents a brief description of the software that was specifically written to analyse studbooks and statistically test results.

1.4.1 Natural History

Husbandry depends on knowledge of a species' natural history. Unfortunately, such information is incomplete for many (animal) species, and often requires long–term field studies to obtain. Some natural history elements can be obtained by analysing studbook data. This is particularly true for life history data such as longevity or reproductive lifespan.

[3]Renamed to Species360 in July 2016.

Chapter 3 presents analyses to retrieve information on lifespan, longevity, reproductive lifespan, litter size, seasonality and inter–birth interval from studbook data. Results of these analyses can be used to adjust husbandry practices, in comparative studies, as baseline data in PHVAs and/or in management of wild populations.

1.4.2 Census

Before computerisation, population growth and birth and death rates in zoo studbook populations were mainly estimated from annual *census counts*, i.e. the number of living individuals at specific dates, as for example published in the International Zoo Yearbook (e.g. Pinder and Barkham 1978). The methodology is similar to wildlife census, except that captive conditions most of the time allow for total counts, and support the collection of additional data, such as on neonatal survival.

Although life table analysis is, in principle, more accurate in estimating population growth, this method does not fully replace census analysis in studbook management. An historical overview of population size, and imports, births, exports and deaths, is essential for understanding the dynamics of the studbook population. Furthermore, census analysis does not require identification of individuals, age of individuals or parentages and can, therefore, be applied to semi–wild or reintroduced populations.

Chapter 4 presents analyses of census counts. Births, deaths and migration (import and export) are considered as *census events* as they are counted between census intervals. Chapter 5 presents analyses of census events to explain trends in population size. In addition, census data on sex–ratio at birth are used to study *sex allocation*. Chapter 6 explores the use of exponential and logistic growth models to describe development of studbook populations.

1.4.3 Life Tables

Future development of a population not only depends on its size, but also, as fecundity and mortality are age related, on its age structure. For example, a population in which most of the individuals are close to the post–reproductive age will decline in the short term, and may never recover. Age structures in males and females are often graphically presented as two opposing vertical histograms, named *age pyramids* (although the shape could be different). Chapter 7 describes the use of age pyramids as a tool to obtain a rough idea about population dynamic processes that occurred in the past and about how the population may develop.

Age–specific fecundity and mortality tables reflect the life history of individuals in the studbook population. These tables can be constructed directly from studbook data. The methods used are derived from those used in wildlife management (e.g.

Caughley 1977). However, the access to dates of birth and death allow the application of refinements such as prorating (Odum and Smith 2001) and censoring of data (Faust et al. 2012). Chapter 7 describes these methods and techniques.

Life tables[4] are often presented in the context of future projections, but are equally important in evaluating husbandry and reproductive management. The significance of life table data depends on the studbook (=sample) size. Populations that have been recently established may not provide sufficient data from which to draw reliable conclusions.

Chapter 8 presents methods to estimate variances and confidence limits from underlying statistical distributions (binomial and Poisson) and bootstrap techniques. In addition, statistical data manipulation such as pooling and smoothing, and curve fitting with parametric models, e.g. Gompertz and Siler curves (Gompertz 1825; Siler 1979), are presented.

Chapter 9 presents techniques to estimate population growth rates, generation time, reproductive value and stable age distribution from fecundity and mortality tables. Future projections are important in assessing the viability of populations and whether (and which) measures need to be taken to increase or to control population size. Chapter 11 describes techniques to project population development from census data and life tables.

Survival analyses based on age classes do not take full advantage of the (usually) accurate dates of birth and death in studbooks. Chapter 10 describes the use of *Kaplan–Meier Product Limit Estimator* (Kaplan and Meier 1958) and *Cox Proportional Hazard Regression model* (Cox 1972). Survival in these models is not estimated at fixed age intervals but at each time a death has occurred in the population. Both survival models were developed for clinical trials, and are extremely useful for comparative studies on factors affecting survival in studbook populations (e.g. Princée and Glatston 2016).

1.4.4 Genetic Analyses

Practical experiences in wildlife management contributed to the development of demographic management of captive populations (Foose and Ballou 1988), whereas applying population genetics to management of real populations had to start more or less "from scratch". The available models in the early 1980s were mostly based on *idealised* populations (Crow and Kimura 1970; Nei 1975), and the genetics used in livestock breeding was very much focussed on selective breeding (e.g. Falconer 1960).

Much of the practical implementation of population genetics in the management of captive populations, such as the use of founder representation, mean kinship values, development of simulation models and group management, can be attributed

[4] A life table in studbook management refers to fecundity and mortality combined.

to "zoo geneticists" (e.g. Ballou and Lacy 1995; Foose 1983; Lacy et al. 1995; Princée 1995). Now, these methods have become part of general conservation genetics and are well described in text books (e.g. Allendorf et al. 2013; Frankham et al. 2010). The genetics section of this book therefore only provides a brief overview of population genetic theory and instead explores what studbook data can contribute to genetic management.

Chapter 12 presents a brief overview of the measures of genetic variation that are used in genetic management of studbook populations. This chapter also pays attention to different interpretations of the term *generation*, and their associated methods. As a result, the calculated generation of individuals can differ between methods. Differences between calculation methods and their practical use are also discussed in Chap. 12.

Chapter 13 briefly describes the process of inbreeding and inbreeding depression due to consanguineous matings and the calculation of inbreeding coefficients. This chapter focusses on the various methods available to detect inbreeding depression from studbook data.

The loss of genetic variation due to genetic drift and the methods to estimate effective population size are covered in Chap. 14. This chapter also discusses the use of computer simulation models that have been developed to estimate loss of genetic variation in real (pedigreed) populations.

Chapter 15 discusses avoidance of selection in captive populations. Founder representation and mean kinship are presented in this chapter in the context of detecting and avoiding potential selection in studbook populations.

Although the importance of maintaining phenotypic variation in zoo populations is recognised (e.g. Arnold 1995), quantitative genetics has not (yet) found its way into studbook management. Chapter 16 explores methods and techniques, as applied to wild populations in ecological studies (e.g. Charmantier et al. 2014), to estimate *heritability* in life history traits from studbook data.

1.5 Conservation Data

Lack of predators, stable food supply and veterinary care represent some of the differences between captive and wild environments. It can be expected that natural history elements such as life history will differ between captive and wild populations. This generates the question of whether the use of data from captive animals in conservation and wildlife management is justified. Chapter 17 discusses the extent to which captive- and wild-sourced data differ; the availability of wild data; and how to handle data from captivity when applied to field conservation.

1.6 More Topics

The quality of studbook data is not considered in the chapters that describe the various demographic and genetic analyses. Although it is generally true that studbook data are more accurate (and detailed) than data collected from wild populations, they are not perfect. For example, a microsatellite DNA study in Przewalski's horses showed that some horses in a particular herd had different parents to those listed in the studbook (Bowling et al. 2003). The age of wild–born individuals is often, depending on the species and method of collection, an estimate or unknown.

Data errors, incomplete and missing data can affect the results of demographic and genetic analyses. Zoo organisations provide guidelines for improving the quality of studbook data (see Thompson et al. 1997; Wilcken and Lees 2012), but it is not always possible to solve all cases of incomplete data. However, that is not necessarily a reason to exclude incomplete data records from analyses. Chapter 18 discusses practical aspects of handling missing data in demographic and genetic analyses.

This book uses statistics as a tool to help in interpreting the results of analyses of often complex (studbook) data. Chapter 19 provides additional information on statistical issues with dependent data, i.e. pseudoreplication (Hurlbert 1984), and incomplete data e.g. estimated dates.

There are likely many more topics on studbooks that are not explored in this book. The criterion for selecting topics was based on availability of data that are stored as standard in studbook databases i.e. parents, sex, dates of birth/death and locations of birth/death. As the various topics in this book illustrate, these minimal data sets already embed a wealth of information that can be used in studbook management and wildlife conservation.

The on–going development of the Species360/ZIMS animal record keeping system will undoubtedly result in extending studbooks with the richness of animal data as stored at the level of individual zoological institutions. It will then be possible to relate demography and genetics to husbandry practices, veterinary medicine or behavioural studies without sending questionnaires to individual institutions. Examples of such studies could include explaining trends in mortality with post mortem data; extending heritability studies to e.g. behaviour, colour, weight and size; and relating reproductive success with group compositions.

References

Ahrens TG (1921) The present status of the European bison or wisent. J Mammal 2(2):58–62
Ahrens TG (1923) Proposals for the preservation of the wisent. Science 58(1493):107
Allendorf FW, Luikart G, Aitken SN (2013) Conservation and the genetics of populations, 2nd edn. Wiley-Blackwell, Chichester
Anonymous (1962) Census of rare mammals in captivity. Int Zoo Yearb 3:302–313

Anonymous (1968) Rare animal studbooks. Int Zoo Yearb 8:382–383
Anonymous (1975) Studbooks and world registers for rare species of wild animals in captivity. Int Zoo Yearb 15:430–434
Anonymous (1980) Studbooks and world registers for rare species of wild animals in captivity. Int Zoo Yearb 20:485–490
Anonymous (2015) International studbooks for rare species of wild animals in captivity. Int Zoo Yearb 49:402–431
Arnold SJ (1995) Monitoring quantitative genetic variation and evolution in captive populations. In: Ballou et al. (1995), pp 295–317
Ballou JD, Lacy RC (1995) Identifying genetically important individuals for management of genetic variation in pedigreed populations. In: Ballou et al. (1995), pp 76–111
Ballou JD, Gilpin M, Foose TJ (eds) (1995) Population management for survival and recovery: analytical methods and strategies in small population conservation. Methods and cases in conservation science series. Columbia University Press, New York
Ballou JD, Lees C, Faust LJ, Long S, Lynch C, Bingaman Lackey L, Foose TJ (2010) Demographic and genetic management of captive populations. In: Kleiman DG, Thompson KV, Kirk Baer C (eds) Wild mammals in captivity: principles and techniques for zoo management, 2nd edn. The University of Chicago Press, Chicago, pp 219–252
Ballou JD, Lacy RC, Pollak JP (2011) PMx: software for demographic and genetic analysis and management of pedigreed populations (version 1.0). Technical report, Chicago Zoological Society, Brookfield
Bouman I, Bouman J (1994) The history of Przewalski's horse. In: Boyd and Houpt (1994), pp 5–38
Bowling AT, Zimmerman W, Ryder O, Penado C, Peto S, Chemnick L, Yasinetskaya N, Zharkikh T (2003) Genetic variation in Przewalski's horses, with special focus on the last wild caught mare, 231 Orlitza III. Cytogenet Genome Res 101:226–234
Boyd L, Houpt KA (eds) (1994) Przewalski's horse: the history and biology of an endangered species. State University of New York Press, Albany
Brewer WH (1893) The mutual relations of science and stock breeding. Science 22(556):169–170
Caughley G (1977) Analyses of vertebrate populations. The Blackburn Press, Caldwell
Charmantier A, Garant D, Kruuk LEB (eds) (2014) Quantitative genetics in the wild. Oxford University Press, Oxford
Coates G (1822) General short–horned herd–book: containing the pedigrees of short–horned bulls, cows, &c. of the improved durham breed. W. Walker, Otley, West Yorkshire
Cox DR (1972) Regression models and life-tables. J R Stat Soc Ser B (Methodol) 34(2):187–220
Crow JF, Kimura M (1970) An introduction to population genetics theory. Harper and Row, New York
Falconer DS (1960) Introduction to quantitative genetics, 1st edn. Longman, London
Faust LJ, Bergstrom YM, Thompson SD, Bier L (2012) PopLink version 2.4. Lincoln Park Zoo, Chicago
Flesness NR (1977) Gene pool conservation and computer analysis. Int Zoo Yearb 17:77–81
Foose TJ (1983) The relevance of captive populations to the conservation of biotic diversity. In: Schonewald-Cox et al. (1983), pp 374–413
Foose TJ, Ballou JD (1988) Population management: theory and practice. Int Zoo Yearb 27:26–41
Frankel OH, Soulé ME (1981) Conservation and evolution. Cambridge University Press, Cambridge
Frankham R, Ballou JD, Briscoe DA (2010) Introduction to conservation genetics, 2nd edn. Cambridge University Press, Cambridge
Glatston AR, Princée FPG (1993) A global masterplan for the breeding of *Ailurus fulgens* in captivity. Technical report, Stichting Koninklijke Rotterdamse Diergaarde (Rotterdam Zoo), Rotterdam
Gompertz B (1825) On the nature of the function expressive of the law of human mortality, and on a new mode of determining the value of life contingencies. Philos Trans R Soc Lond 115:513–583

Hurlbert SH (1984) Pseudoreplication and the design of ecological field experiments. Ecol Monogr 54(2):187–211
Hurst CC (1906) On the inheritance of coat colour in horses. Proc Royal Soc Lond B: Biol Sci 77(519):388–394
Kaplan EL, Meier P (1958) Nonparametric estimation from incomplete observations. J Am Stat Assoc 53:457–481
Lacy RC, Ballou JD, Princée F, Starfield A, Thompson EA (1995) Pedigree analysis for population management. In: Ballou et al. (1995), pp 57–75
Lees C, Wilcken J (2009) Sustaining the ark: the challenges faced by zoos in maintaining viable populations. Int Zoo Yearb 43:6–18
Little CC (1914) Coat color in pointer dogs: examination of stud book from Mendelian viewpoint suggests that relation of brown to black is the same in dogs as in other small mammals which have been investigated, and that there are two types of yellow–practical breeding rules. J Hered 5(6):244–248
McLeod WT (ed) (1987) The Collins dictionary and thesaurus in one volume. Collins, London
Mohr E (1971) The asiatic wild horse, 2nd edn. (English trans: Goodall DM). J. A. Allen, London
Nei M (1975) Molecular population genetics and evolution. North-Holland Publishing Company, Amsterdam
Odum RA, Smith BR (2001) The effects of prorating risk in the development of life-tables. Zoo Biol 20:279–291
Pearson K, Lee A, Bramley–Moore L (1899) Mathematical contributions to the theory of evolution. vi. genetic (reproductive) selection: inheritance of fertility in man, and of fecundity in thoroughbred racehorses. Philos Trans R Soc Lond A: Math Phys Eng Sci 192:257–330
Pinder N, Barkham J (1978) An assessment of the contribution of captive breeding to the conservation of rare mammals. Biol Conserv 13(3):187–245
Princée FPG (1995) Overcoming the constraints of social structure and incomplete pedigree data through low-intensity genetic management. In: Ballou et al. (1995), pp 124–154
Princée FPG, Glatston AR (1992) The red panda *Ailurus fulgens*, as an example of a global animal survival plan. In: Brouwer K, Smits S, de Boer LEM (eds) EEP Yearbook 1991/1992 including the Proceedings of the 9th EEP Conference, Edinburgh, 6–8 July 1992. European Association of Zoos and Aquaria, EAZA/EEP Executive Office, Amsterdam, p 211
Princée FPG, Glatston AR (2016) Influence of climate on the survivorship of neonatal red pandas in captivity. Zoo Biol 35(2):104–110
Ralls K, Brugger K, Ballou J (1979) Inbreeding and juvenile mortality in small populations of ungulates. Science 206:1101–1103
Ricklefs RE (2010) Life-history connections to rates of aging in terrestrial vertebrates. Proc Natl Acad Sci 107(22):10314–10319
Schonewald-Cox CM, Chambers SM, MacBryde B, Thomas L (eds) (1983) Genetics and conservation: a reference for managing wild animal and plant populations. Biological conservation. Benjamin/Cummings, Menlo Park
Siler W (1979) A competing-risk model for animal mortality. Ecology 60(4):750–757
Soulé ME, Gilpin M, Conway W, Foose T (1986) The millenium ark: how long a voyage, how many staterooms, how many passengers. Zoo Biol 5:101–113
Thompson SD, Earnhardt JM, Turner-Erfort G (1997) Guidelines for data entry and maintenance of North American regional studbooks. Lincoln Park Zoo, Chicago
Tong EH (1957) World register of Père David's deer. Proc Zool Soc Lond 129(3):343–449
Volf J (1994) The studbook. In: Boyd and Houpt (1994), pp 61–73
von der Groeben G (1932) Das Zuchtbuch. Berichte der Internationalen Gesellschaft zur Erhaltung des Wisents 5(1):5–50
WAZA (2016) Global Species Management Plans. Technical report, WAZA. www.waza.org
Weatherby J (1791) An introduction to the general stud–book. H. Reynell, London
Wijnstekers W (2011) The evolution of CITES: a reference to the Convention on International Trade in Endangered Species of Wild Fauna and Flora, 9th edn. International Council for Game and Wildlife Conservation, Budapest

Wilcken J, Lees C (eds) (2012) Managing zoo populations: compiling and analysing studbook data, 3rd edn. Australasian Regional Association of Zoological Parks and Aquaria. http://training.isis.org/CommunityLibrary/ManagingZooPopulations-2012edition.pdf

Young AM, Hobson EA, Bingaman Lackey L, Wright TF (2012) Survival on the ark: life–history trends in captive parrots. Anim Conserv 15(1):28–43

Chapter 2
Studbooks and Software

Abstract Studbooks on the red crowned crane, African wild dog, red panda, wolverine, snow leopard, Californian sea lion, blesbok and mountain bongo are used as examples in this book. Short profiles of the species and their studbooks are presented. These include information on geographic distribution and IUCN Red Data List status, scope of the studbook, i.e. global or regional, type of population management and studbook keeper. A brief history of the development of software tools for population management by the international zoo community is presented. This includes computerised animal record keeping systems, such as ARKS and SPARKS, and demographic and genetic analyses that are nowadays integrated in the standard zoo population management software PMx. The Population Management Library (PML) software that was developed specifically for this book is described. This software imports data directly from SPARKS studbooks and applies its own data filters to create subsets for comparative studies. The PML program is used throughout the book for natural history, demographic and genetic analyses of the studbook examples. PML includes basic statistics but uses R through the add-on module studbookR for statistical analyses and graphic representation of population analyses. An overview of the available analyses and statistical tests in PML and studbookR is provided.

2.1 Introduction

The international zoo community had established 190 international and 1,350 regional studbooks on endangered animal species by 2011 (Anonymous 2015). The studbook examples for this book were selected on the basis of personal experience with the species. These studbooks either were the source of topics in this book and/or matched the characteristics to illustrate topics that I had in mind. The next section describes the studbooks that are used throughout this book.

Analysis of studbook data can be tedious or almost impossible to conduct without the aid of computers and dedicated software. The second section of this chapter provides a brief historic overview of (zoo) studbook software and introduces the programs that are used throughout this book.

F.P.G. Princée, *Exploring Studbooks for Wildlife Management and Conservation*, Topics in Biodiversity and Conservation 17, DOI 10.1007/978-3-319-50032-4_2

2.2 Studbook Examples

The results of analyses presented in this book are for illustrative purposes only and should not be used to evaluate population management of the species involved or used as sources for comparative studies. Moreover, older versions of studbooks that are used do not necessarily reflect current management.

The original data of all studbooks are computerised in the program SPARKS v1.66 (Scobie and Bingaman Lackey 1997–2012). Table 2.1 presents a summary of species information as included in each individual studbook database.

2.2.1 *Red Crowned Crane* Grus japonensis

The total wild population of red crowned cranes in south–eastern Russia, north–east China, Mongolia and eastern Hokkaido (Japan) is estimated to be 2,750 individuals, of which 1,650 are mature (BirdLife International 2013). The species is listed as Endangered (EN) in the IUCN Red List of Threatened Species. The European zoo population of red crowned cranes has been managed as an EEP since 1991. The studbook data that are used in this book are current through 1 January 2012 and were compiled by R. Belterman (Rotterdam Zoo, The Netherlands).

Red crowned cranes are monogamous with stable pair bonding for years in the wild (Klenova et al. 2009). This species is held, in accordance with its natural

Table 2.1 Overview of studbooks that are used as examples in this book. Scope of the studbook are EEP: EAZA European Endangered Species Programme, ESB: EAZA European studbook, intl: International studbook

English name	Scope	Type	Incubation Gestation[a]	Date range[b]	N	Update
Red crowned crane	EEP	Egg–laying	31	2	1,170	01 Jan 2012
African wild dog	EEP	Live–bearer	98	3	2,174	01 Mar 2006
					2,697	25 Aug 2013[c]
Red panda	intl	Live–bearer	132	3	2,907	31 Dec 2008
					3,559	31 Dec 2012[c]
Wolverine	EEP	Live–bearer	30	3	159	01 Aug 2001
Snow leopard	intl	Live–bearer	96	3	2,481	17 Mar 2003
Californian sea lion	ESB	Live–bearer	365	3	1,531	31 Dec 2009
Blesbok	ESB	Live–bearer	240	3	1,460	18 May 2009
Mountain bongo	intl	Live–bearer	270	3	2,305	31 Dec 2010

N Total number of individuals in studbook
[a] Incubation time or gestation length in days
[b] Range in birth or hatch dates of same litter or clutch in days
[c] Studbook data used in genetic analyses

breeding behaviour, in (breeding) pairs (Mirande et al. 1996). The number of captive–born red crowned cranes with unknown parentages is therefore relatively low i.e. $\approx$4 %.

2.2.2 *African Wild Dog* Lycaon pictus

The African wild dog has disappeared from much of its former (pan–African) range. The species is virtually eradicated from North and West Africa, and greatly reduced in Central Africa and north–east Africa. The largest populations remain in southern Africa and the southern part of East Africa (Woodroffe and Sillero–Zubiri 2012). Dispersal distances of up to 520 km have been observed in African wild dogs (Masenga et al. 2016). This can explain new observations, such as in the southeastern part of the Central African Republic (Hickisch and Aebischer 2013).

The total wild population is estimated to be 6,600 adults and yearlings. The number of sexually mature individuals is 1,400 as only alpha males and females and sub–dominant animals that breed successfully are included. The African wild dog is listed as Endangered (EN) in the IUCN Red List (Woodroffe and Sillero–Zubiri 2012).

The African wild dog is a social species living in packs with on average 10 adults and yearlings, although larger packs have been observed (Alderton 1994; Courchamp and Macdonald 2001). The alpha male and female are the main breeders in the pack, with their same sex siblings as helpers (Girman et al. 1997). This composition is followed in population management programmes such as in the EEP for African wild dogs.

An international studbook for African wild dogs was established in 1972, followed by regional breeding programmes. The studbook data of the EEP programme compiled by W. and H. Verberkmoes (GaiaZoo, Kerkrade, The Netherlands), current through 1 March 2006, is used. A newer version, current through 25 August 2013, and compiled by R. Barnes (Port Lympne Reserve, Lympne, United Kingdom) is used for quantitative genetics.

2.2.3 *Red Panda* Ailurus fulgens

The red panda lives in temperate cloud forests of the Himalayan region and western China. Two subspecies are described: the Nepalese red panda (*Ailurus fulgens fulgens*) with an estimated population of 600–1,000 and the Chinese red panda (*A. f. styani*) with an estimated population of 3,000 7,000. The species is listed as Endangered (EN) in the IUCN Red List (Glatston et al. 2015).

An international studbook including both red panda subspecies was initiated in 1978 and since then maintained by Rotterdam Zoo (The Netherlands).

The first edition of the International Red Panda Studbook already included demographic analyses (Glatston 1980). The Nepalese red panda population belongs to the first examples of pedigree analysis using computer simulation models (Princée 1988, 1989a). The zoo populations of both red panda subspecies are managed through regional management programmes. Since 2015 these regional programmes cooperate as a *Global Species Management Plan (GSMP)* (WAZA 2016).

Two studbook data sets, one current through 31 December 2008 and a more recent set current through 31 December 2012, were used in this book. These data were compiled by international studbook keeper A. Glatston (Rotterdam Zoo). The 2012 dataset was used for genetic analyses.

2.2.4 *European Wolverine* **Gulo gulo gulo**

The wolverine is listed as Least Concern (LC) in the IUCN Red List at the species level, with large populations in North America and northern Asia. The European wolverine has an estimated wild population of 2,260 and is considered Vulnerable (VU) (Abramov et al. 2009). The EEP programme for the European wolverine was established in 1994 and since then the studbook has been maintained by the Nordic Ark (Hunnebostrand, Sweden). The computerised data set which is used in this book is current through 1 August 2001 and was compiled by EEP coordinator L. Blomqvist. This early version of the wolverine studbook has only 159 records and is mainly used to illustrate the challenges of demographic analysis with small sample sizes.

2.2.5 *Snow Leopard* **Uncia uncia**

The natural distribution of the snow leopard is restricted to the high mountains of Central Asia. The wild population is estimated to be between 4,500 and 6,500 individuals and the species is listed as Endangered (EN) in the IUCN Red List (Jackson et al. 2008).

The International Pedigree Book for snow leopard was established in 1976 (Blomqvist 2004). The computerised data set used in this book is current through 17 March 2003, and was compiled by international studbook keeper L. Blomqvist (Helsinki Zoo, Finland). This particular data set comprises records on 2,500 individuals with only a single captive–born individual with unknown parents. The size and complete pedigree make this studbook a good example for comparative studies.

2.2.6 *Californian Sea Lion* Zalophus californianus

The Californian sea lion is classified as Least Concern (LC) in the IUCN Red List (wild population ≈ 390,000) (Aurioles–Gamboa and Hernández–Camacho 2015). However, this species is protected under the U.S. Marine Mammal Protection Act of 1972. This means that individual sea lions need to be registered. The European studbook compiled by G. H. Meijer (Ouwehand Zoo, Rhenen, The Netherlands) with data current through 31 December 2009 is used here. This studbook evolved from a Dutch national studbook that was established in the early 1980s. Californian sea lions show strong seasonality in births (Temte 1993) and are used as an example of applying circular statistics to test whether stillbirths are premature births.

2.2.7 *Bongo* Tragelaphus eurycerus

The IUCN recognises two subspecies of the bongo antelope: the lowland or western bongo (*Tragelaphus eucerycerus eurycerus*) in the lowland rainforests of West and Central Africa, and the eastern or mountain bongo (*T. e. isaaci*) in montane forests of Kenya (Aberdares, Mount Kenya, Mau Forest and Eburu Forest). The lowland bongo is classified as Near Threatened (NT) in the IUCN Red List (wild population ≈ 28,000 in 1999) due to ongoing population decline as a result of habitat loss and hunting pressure. The eastern bongo, with an estimated population size of 74–140 individuals, is classified as Critically Endangered (CR) (IUCN SSC Antelope Specialist Group 2008).

Although the bongo is the largest forest antelope, with males weighing up to 400 kg (Estes 2012), its secretive lifestyle means that knowledge of natural history on this species is limited. A *Population Viability Analysis (PVA)* of lowland bongo in Dzanga Sangha Protected Areas (Central African Republic) (Princée 2011) used life history data extracted from the International studbook for the eastern/mountain bongo. This studbook contains 2,305 records and was compiled by L. Bosley with data current through 31 December 2009 (Bosley 2010). The PVA study on lowland bongo is used as an example in Chap. 17 to illustrate the use of data from captivity in wildlife conservation.

2.2.8 *Blesbok* Damalisicus pygargus phillipsi

The original distribution of blesbok is open grassland in southern Transvaal, Orange Free State and northeastern Karoo (South Africa). This species was almost driven to extinction in the nineteenth century due to excessive hunting for its hide by

European settlers (du Plessis 1972). The population recovered from some 2,000 individuals, mainly kept on private farmland, to an estimated 235,000–240,000 in 1998 (East 1999). Blesbok are listed as Least Concern (LC) in the IUCN Red List (Lloyd and David 2008). A European studbook was established in 1998. Studbook data compiled by H. Verberkmoes (GaiaZoo, Kerkrade, The Netherlands) and current through May 2009 are used as a case study in this book.

This studbook characterises zoo species that are kept in (harem) groups and have a history of successful reproduction long before the principles of maintenance of genetic variation were introduced. Some 30 % of individuals listed in the European studbook have either a single or both parents unknown. The majority of these individuals were born before the European studbook was established in 1998. Missing parentages affect both demographic analyses and genetic analyses, and thus, management. Census and census-style analyses are applied to determine population trends in the studbook population of the blesbok, as if it was a survey on the wild population.

2.3 Software Tools

The introduction of personal computers in the mid 1980s has been tremendously beneficial in evolving the science behind the management of small populations. Database systems developed by the *International Species Information System (ISIS)* allowed zoological gardens more easily to manage in–house data, from daily observations to births, deaths and transfers, in the program *Animal Record Keeping System (ARKS)*. Exchange of data on individuals between zoological collections was facilitated through the central ISIS on–line database. Meanwhile, ARKS has been replaced by the new *Zoological Information Management System (ZIMS)* and internet has replaced floppy disks to update the central database.

Development of studbook software followed shortly after the introduction of ARKS. The dBase programmes "Omaha" and "Houston" were the predecessors of *Zooresearch Studbook Management (ZRBOOK)* (Princée 1989b) and the *Single Population Animal Record Keeping System (SPARKS)* (Scobie and Flesness 1989). The ZRBOOK software was mainly used in Europe and was discontinued in favour of further development of the global studbook program SPARKS. This DOS program is still in use and has been updated to export studbook data to analysis programs (Scobie and Bingaman Lackey 1997–2012). The studbook program *PopLink* is an alternative to SPARKS, as it runs under the Windows operating system (Faust et al. 2012).

A large variety of demographic and genetic analyses have been developed since electronic studbook databases became available (see overview in Ballou et al. 2010). Chapter 14 describes genetic simulation models that use studbook data. Most of the stand–alone analyses have been re-designed and grouped into the single population management program *PMx* (Ballou et al. 2011).

One impetus for this book was the need to design new analysis software to tackle biological topics that were not covered in other software. The next section briefly describes this software.

2.3.1 Population Management Library

The *Population Management Library (PML)* software (Princée 2014a) is a set of library modules for demographic and genetic analyses of studbook data. It includes a command–line interface which uses its own script language to invoke analyses and to apply population view settings (data filters). The software is written in C++ and is designed to be cross–platform (Mac OS X, Linux/Unix and Windows). It is available under an open source license at "http://www.princee.com".

Table 2.2 Analyses and statistical tests in the Population Management Library/studbookR

Analysis	Description
Age at death/longevity	Summary statistics[a]; histogram
Age at reproduction	Summary statistics; histogram
Clutch/litter size	Summary statistics; histogram
Inter–birth interval	Summary statistics; histogram
Birth/litter season	Mean date; rose diagram; circular statistics; effects of latitude, captive generation and age of dam
Death season	Mean date; rose diagram; circular statistics
Annual census	Graphics; annual growth rates; logistic and exponential model
Births	Graphics; birth rates, runs test on sex–ratio
Deaths	Graphics; death rates
Neonatal deaths	Graphics; death rates
Imports	Graphics; import rates
Exports	Graphics; export rates
Age distribution	Age pyramid
Fecundity	Age–class specific fecundity rates; confidence limits
Mortality	Age–class specific mortality rates; confidence limits
Full life table	Reproductive values; stable age distribution; $R0$, r, λ
Survival analyses	Kaplan–Meier product estimator; Cox's proportional hazard regression; comparative studies on e.g. sex, inbreeding, captive generation, location and climate (if lat/lon data are available)
Age at first breeding	Kaplan–Meier product estimator
Founder representation	In group(s) and individuals
Inbreeding depression	Survival curves and lethal equivalents
Heritability	Parent–offspring regression and animal model on natural history "traits" in studbook data

[a] Summary statistics includes Tukey's five number summary, mean and variance

The base of the PML is formed by a redesign of analyses that were part of the ZRBOOK suite (Princée 1991). The software not only includes new analyses but also implements basic statistical tests and the resampling techniques *bootstrap* and *jackknife* (Efron 1979; Quenouille 1949; Tukey 1958). PML can import studbooks that have been created in SPARKS.[1]

A large variety of view filters can be applied to the data, including sex, (census) dates, locations, inbreeding levels, generation number, neonatal age and stillbirths and accuracy level of dates. These filters are used to define the scope of the analyses for management or comparative studies.

The statistics program *R* (R Core Team 2015) and add–on packages are used for additional (more elaborate) statistical tests and graphic capabilities. PML exports analysis results and/or filtered studbook data which can then be imported in R (and in most spreadsheet programs). The R script files that import and further analyse these files are bundled in the package *studbookR* (Princée 2014b), which is part of the PML distribution.

Table 2.2 presents an overview of the available analyses and statistical tests in PML and/or studbookR. The R add–on packages that are used within studbookR are mentioned and cited in the relevant chapters. An overview of available script commands in the PML library and functions in the studbookR package are provided in Appendix A.

References

Abramov A, Belant J, Wozencraft C (2009) *Gulo gulo*. The IUCN red list of threatened species 2009: eT9561A13000870. doi:10.2305/IUCN.UK.2009-2.RLTS.T9561A13000870.en

Alderton D (1994) Foxes, wolves and wild dogs of the world. Blandford, London

Anonymous (2015) International studbooks for rare species of wild animals in captivity. Int Zoo Yearb 49:402–431

Aurioles–Gamboa D, Hernández–Camacho J (2015) *Zalophus californianus*. The IUCN red list of threatened species 2015: eT41666A45230310. doi:10.2305/IUCN.UK.2015-4.RLTS.T41666A45230310.en

Ballou JD, Lees C, Faust LJ, Long S, Lynch C, Bingaman Lackey L, Foose TJ (2010) Demographic and genetic management of captive populations. In: Kleiman DG, Thompson KV, Kirk Baer C (eds) Wild mammals in captivity: principles and techniques for zoo management, 2nd edn. The University of Chicago Press, Chicago, pp 219–252

Ballou JD, Lacy RC, Pollak JP (2011) PMx: software for demographic and genetic analysis and management of pedigreed populations (version 1.0). Technical report, Chicago Zoological Society, Brookfield

BirdLife International (2013) *Grus japonensis*. The IUCN red list of threatened species 2013: eT22692167A48057181. doi:10.2305/IUCN.UK.2013-2.RLTS.T22692167A48057181.en

Blomqvist L (2004) International pedigree book for snow leopards (*Uncia uncia*), vol 8. Helsinki Zoo, Helsinki

[1]SPARKS studbooks are converted from dBase format to SQLite3.

Bosley LF (2010) International studbook for Eastern/Mountain Bongo (*Tragelaphus eurycerus isaaci*) – year 2009 historical edition, vol XXIV. Logsden

Courchamp F, Macdonald DW (2001) Crucial importance of pack size in the African wild dog *Lycaon pictus*. Anim Conserv 4:169–174

du Plessis S (1972) Ecology of blesbok with special reference to productivity. Wildl Monogr 30:3–70

East R (1999) African antelope database 1998. IUCN, Gland/Cambridge

Efron B (1979) Bootstrap methods: another look at the jackknife. Ann Stat 7(1):1–26

Estes RD (2012) The behavior guide to African mammals: including hoofed mammals, carnivores, primates, 20th edn. University of California Press, Berkeley

Faust LJ, Bergstrom YM, Thompson SD, Bier L (2012) PopLink version 2.4. Lincoln Park Zoo, Chicago

Girman DJ, Mills MGL, Geffen E, Wayne RK (1997) A molecular genetic analysis of social structure, dispersal, and interpack relationships of the African wild dog (*Lycaon pictus*). Behav Ecol Sociobiol 40:187–198

Glatston AR (1980) The red or lesser panda studbook, number 1. Stichting Koninklijke Rotterdamse Diergaarde, Rotterdam

Glatston AR, Wei F, Zaw ZT, Sherpa A (2015) *Ailurus fulgens*. The IUCN red list of threatened species 2015: eT714A45195924. doi:10.2305/IUCN.UK.2015-4.RLTS.T714A45195924.en

Hickisch R, Aebischer T (2013) Evidence of African wild dogs in the Central African Republic. Canid Biol Conserv 16(3):8–11

IUCN SSC Antelope Specialist Group (2008) *Tragelaphus eurycerus*. The IUCN red list of threatened species 2008: eT22047A9351710. doi:10.2305/IUCN.UK.2008.RLTS.T22047A9351710.en

Jackson R, Mallon D, McCarthy T, Chundaway RA, Habib B (2008) *Panthera uncia*. The IUCN red list of threatened species 2008: eT22732A9381126. doi:10.2305/IUCN.UK.2008.RLTS.T22732A9381126.en

Klenova AV, Volodin HA, Volodina EV (2009) Examination of pair-duet stability to promote long-term monitoring of the endangered red-crowned crane (*Grus japonensis*). J Ethol 27(3):401–406

Lloyd P, David J (2008) *Damaliscus pygargus*. The IUCN red list of threatened species 2008: eT30208A9530977. doi:10.2305/IUCN.UK.2008.RLTS.T30208A9530977.en

Masenga EH, Jackson CR, Mjingo EE, Jacobson A, Riggio J, Lyamuya RD, Fyumagwa RD, Borner M, Røskaft E (2016) Insights into long–distance dispersal by African wild dogs in East Africa. Afr J Ecol 54(1):95–98

Mirande CM, Gee GF, Burke A, Whitlock P (1996) Egg and semen production. In: Ellis DH, Gee GF, Mirande CM (eds) Cranes: their biology, husbandry and conservation. U.S. Department of the Interior, National Biological Service, International Crane Foundation, Washington, DC/Baraboo, pp 45–57

Princée FPG (1988) Genetic variation in the zoo population of the red panda subspecies *Ailurus fulgens fulgens*. Zoo Biol 7(3):219–231

Princée FPG (1989a) Preservation of genetic variation in the red panda population. In: Glatston AR (ed) Red panda biology, SPB Academic Publishing, The Hague, pp 171–182

Princée FPG (1989b) Zooresearch studbook management program version 1.03, software and manual. National Foundation for Research in Zoological Gardens, Amsterdam

Princée FPG (1991) Zooresearch studbook management program version 1.04, software and manual. National Foundation for Research in Zoological Gardens, Amsterdam

Princée FPG (2011) Population viability assessment western/lowland bongo in Dzanga–Sangha Protected Areas. Technical report, WWF-CAR, Bayanga, Central African Republic

Princée FPG (2014a) Population Management Library (PML) version 1.0, software and manual. East Rudham. http://www.princee.com

Princée FPG (2014b) studbookR: statistical analyses of studbooks version 0.2, software and manual. East Rudham. http://www.princee.com

Quenouille MH (1949) Approximate test of correlation in time–series. J R Stat Soc Ser B (Methodol) 11(2):68–84

R Core Team (2015) R: a language and environment for statistical computing. R Foundation for Statistical Computing, Vienna. https://www.R-project.org/

Scobie P, Bingaman Lackey L (1997–2012) Single Population Animal Record Keeping System (SPARKS) versions 1.42–1.66. International Species Information System, Eagan

Scobie P, Flesness NR (1989) Single Population Animal Record Keeping System (SPARKS), software and manual. International Species Information System, Apple Valley

Temte JL (1993) Latitudinal variation in the birth timing of captive California sea lions and other captive North Pacific pinnipeds. Fish Bull 91(4):710–717

Tukey JW (1958) Bias and confidence in not–quite large samples. (Abstract of preliminary report). Ann Math Stat 29(2):614

WAZA (2016) Global Species Management Plans. Technical report, WAZA. http://www.waza.org

Woodroffe R, Sillero–Zubiri C (2012) *Lycaon pictus*. The IUCN red list of threatened species 2012: eT12436A16711116. doi:10.2305/IUCN.UK.2012.RLTS.T12436A16711116.en

Part II
Demographic Analyses

Chapter 3
Natural History

Abstract Natural history data are important for husbandry and population management of wild species in captivity, although these data are often incomplete or missing on wild species. However, some of these data can be extracted from studbooks. They are referred to as natural history *elements* in this book to distinguish them from data obtained from the wild. Histograms and descriptive statistics such as Tukey five number summary are used to describe characteristics of (reproductive) lifespan, litter/clutch size and inter–birth interval in the population. Results from these statistics are placed in the context of husbandry and population management. Techniques to detect outliers (possible artefacts) in longevity, first age at reproduction and maximum lifespan are presented. The non–parametric Wilcoxon two–sample test is applied to test for differences in (reproductive) lifespan between sexes and different periods of time. The problem of assessing litter size in species that use dens, and a possible solution, is discussed in this chapter. Circular statistics are applied in seasonality analyses of births and deaths. Effects of latitude (northern and southern hemispheres) on birth season are discussed. Additional topics such as the use of seasonality analysis in detecting premature births and sex–bias are explored.

3.1 Introduction

Population management starts with having basic knowledge of the natural history of the species involved. Natural history refers to life history information such as longevity and reproductive lifespan, to type of reproduction (sexual or asexual), method of reproduction (viviparous, oviparous or ovoviviparous) and social mating systems (pair, harem, multi–male/female groups). In short, the entire biology of a species.

Natural history in general refers to observations of wild populations. Unfortunately, these data are incomplete for many (endangered) species. A secretive lifestyle or inaccessible habitat makes it difficult to obtain data from the wild, or long–term studies are required to collect sufficient data. In addition, it is harder to study species that are rare. It is not uncommon for zoo data, e.g. on longevity,

F.P.G. Princée, *Exploring Studbooks for Wildlife Management and Conservation*, Topics in Biodiversity and Conservation 17, DOI 10.1007/978-3-319-50032-4_3

gestation length or litter size, to be used to characterise species in encyclopaedias and field guides (e.g. Estes 2012; Kingdon 2015). The term *natural history elements* is used throughout this book and refers to data obtained from captive populations.

Knowledge of natural history is the basis from which to manage populations. It provides the information source, such as appropriate diet or social structure, for husbandry (keeping animals alive and healthy). These data can be considered as a foundation before even thinking about population management. Studbooks contain a wealth of data that are part of a species' natural history. In particular, the older (larger) studbooks provide sample sizes that allow for the application of statistical tests. But the issue is often "how to extract these data?". Well, software programs like SPARKS (Scobie and Bingaman Lackey 1997–2012), PopLink (Faust et al. 2012) and PML (Princée 2014) (this last program is used in this book), provide analyses that can reveal natural history elements. Yet population management in zoos often tends to be focussed on future projections with respect to population growth and maintenance of genetic variation, "forgetting" that other valuable data (and analysis tools) are available.

Natural history data are not only important for managing zoo populations. Conservation models like *Population Viability Analysis (PVA)* require data such as litter/clutch size, inter–birth interval and reproductive lifespan. Zoo populations are the source whenever these data are incomplete or missing on wild populations. These are the data that can be used to understand and, ultimately, to protect species in their wild habitat.

This chapter introduces some basic analyses to extract information on natural history elements, such as longevity, life expectancy, reproductive lifespan, litter size, inter–birth interval and seasonality patterns, from studbook data. The emphasis of this chapter is on exploring the data set and interpreting the results, rather than solely learning the methodologies (and statistics). Descriptive statistics, comparable to *Tukey's five number summary* (Tukey 1977), are used to provide first impressions of the data.

More elaborate analyses are available in some cases, such as age–specific life table analyses, which will be discussed in later chapters (see Chap. 7). The International studbook for snow leopards (Blomqvist 2004) is used as the main example in this chapter.

3.2 Lifespan and Longevity

The lifespan of an individual is the interval between its birth and its death, or more simply the age at death. Lifespan differs between individuals as some die shortly after birth, while others survive even beyond the age of reproduction. Longevity (symbol: Ω) is not strictly defined in demography (Barlow and Boveng 1991), but generally refers to the longest lived individual as observed in a species or in a specific population (see Sect. 3.2.2). Analyses of ages at death are not limited to determining longevity. Results of these analyses can, for example, pinpoint age

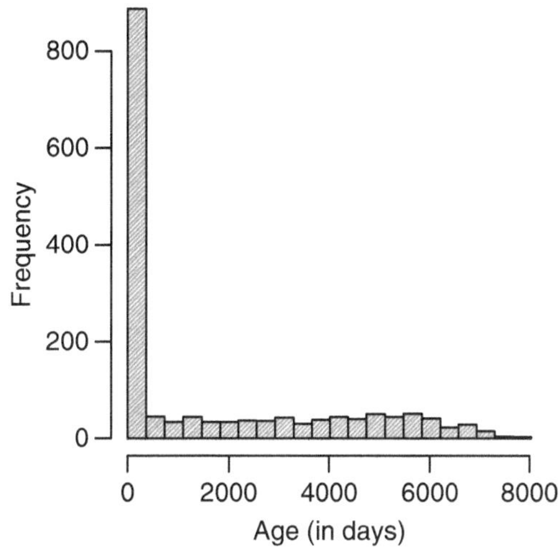

Fig. 3.1 Frequency counts of age at death in captive snow leopards from 1890 to 2003 ($n = 1{,}606$). Accurate data only. Bars correspond to 1 year intervals

Table 3.1 Summary statistics of age at death (in days) in captive snow leopards for different levels of date precision from 1890 to 2003. See text for explanation of precision categories

Precision	1st	Median	Mean	3rd	Maximum	n
Accurate	4	142	1,694	3,372	7,981	1,606
By month[a]	4	151	1,697	3,374	7,981	1,609
By year[a]	5	160	1,722	3,480	7,981	1,640
All[a]	7	365	1,782	3,510	7,981	1,818

1st first quartile, *3rd* third quartile, *n* sample size
[a] Includes data of previous category/categories

groups with high mortality or indicate *mean or median life expectancy*, which is the mean or median age a newborn is expected to reach. Analyses of age at death should initially include individuals of all sexes,[1] restricted to individuals with accurate dates of *both* birth and death.

Graphic representations are important as they provide a "feeling" for the distribution pattern of the data and, therefore, for the interpretation of statistics in terms of biology. Figure 3.1 presents a histogram with frequency counts of the age at death, grouped in age classes of 1 year, in captive snow leopards. The summary statistics of the same data (group *Dead*) – but based on age in days – is presented in Table 3.1.

High mortality in the first year is the most obvious feature that can be observed in Fig. 3.1. High juvenile mortality is common in wild populations of many species (e.g. Caughley 1966). No deaths have been recorded after age class 21, implying that longevity is between 21 and 22 years. The results of summary

[1]The following sex–group categories are included: male, female, unknown sex, hermaphrodite and abnormal sex.

statistics (Table 3.1) on this data set provide numerical "support" for the features observed in Fig. 3.1. Note that the data in these statistics were not grouped in age classes. Fifty percent (the median) of deaths occur between 0 and 142 days ($\approx 4\frac{1}{2}$ months), while the maximum observed age at death in this population is 7,981 days ($\approx$22 years).

3.2.1 *Estimated Dates*

Exact dates of birth are not always available. This is especially the case for wild–caught individuals whose age (i.e. dates of birth) needs to be assessed from their estimated age at arrival. Such dates will almost always, by definition, be estimates. However, such information may not be available for individuals that were imported in the early phase of a studbook. Dates of birth will consequently remain unknown. Dates of death can also be estimates or unknown, for a variety of reasons.

The SPARKS studbook software provides the option of entering the precision of dates (Scobie and Bingaman Lackey 1997–2012). The following categories are used: accurate, estimated by day, month or year, date range and unknown date. Estimated by day refers to a precision of a few days, while estimated by month or year refer to a date within the given month or year, respectively. In general, the midpoint of a month (day 15/16) or a year (1 July) are used for these estimates. Date ranges are provided in months or in years. In the case that the date of birth is completely unknown, it equals the date of first of arrival in the studbook population.

Incomplete dates, and especially unknown dates, can have effects on demographic analyses. These effects depend on the precision of estimates and the proportion of data that are estimates. Handling of estimates and missing data will be discussed in more detail in Chap. 18. Nevertheless, lifespan analyses on accurate data sets can provide a guide to what extent estimates can comfortably be included.

Longevity of 7,981 days, based on accurate dates, in snow leopards does not differ from values estimated in data sets that include estimated dates of birth by day or month (Table 3.1). This means that including individuals with estimated dates by day and month would have negligible impact in assessing longevity. Estimates by year could introduce an error of less than 5 % on determining longevity in this species. The level of precision depends on the resolution of an analysis. For example, individuals with estimated dates by day or month should not be included in analyses of perinatal or neonatal mortality.

Effects of including estimated dates of birth and/or death on lifespan analyses are presented in Table 3.1. The international studbook on the snow leopard (Blomqvist 2004) does not include records with dates that are estimates by day. Each category also includes records of higher precision. For example, "estimated by year" includes accurate dates and estimates by month. The category 'all' includes accurate dates, estimates and unknown dates.

The maximum observed ages (longevity) in the four data sets do not differ, indicating that this value is based on a record with accurate dates. The general trend is that the quartiles and the mean values increase with lower accuracy levels (Table 3.1). This can be explained by the fact that individuals with estimated dates of birth have, in general, survived the first week or months, before being registered. Snow leopards with unknown dates of birth are likely to have passed the juvenile stage. This is reflected by the median age at death in the "all" group. This value has more than doubled compared to the median in the data set that includes estimates (Table 3.1).

It is recommended to inspect studbook data on the origin of individuals with estimated or unknown dates. For example, captive–born snow leopards are all registered with accurate dates of birth and death. This means that estimated data refer to wild–caught individuals. Therefore, further analyses in this section will be based on accurate dates only i.e. on captive–born individuals.

3.2.2 Longevity

Longevity records are popular in zoo literature (e.g. Brouwer et al. 1994; Jones 1993). This information served, especially before the introduction of computerised studbooks, as an indication of success in keeping wild animals (alive). However, one should realise that the (general) definition of longevity means that its value is based on a single observation – the age of the longest lived individual. This single record may not be representative for the species, e.g. when analyses are limited to a single (sub–)population. Since longevity is only known at death, living individuals that are older than previous (dead) ones are not "counted". For example, this has been observed in African elephants (*Loxondonta africana*) and Asian elephants (*Elephas maximus*) in European and North American zoo populations (Wiese and Willis 2004), where living elephants are older than the previous longevity record (in these populations).

3.2.2.1 Artefacts

The oldest observed ages at death in studbooks could be *artefacts* due to errors in recording dates of birth and/or death. Artefacts can also be introduced by veterinary intervention in cases where the life of older individuals is "artificially" prolonged. It is clear that such artefacts should ideally be excluded from analyses, especially in the context of gaining information on natural history.

Statistical methods such as the Dixon's test and Grubbs' test (Komsta 2011) can be used to detect artefacts that show up as *outliers* in data sets. However, these tests assume that data are normally distributed (Sokal and Rohlf 2012). Figure 3.1 shows that the age at death is extremely skewed to the left (first year) and, therefore, cannot be considered to be normally distributed.[2] This means that outliers may not be detected or may be wrongly detected. Therefore, these tests should be applied with caution.

Studbooks are usually small data sets which do not necessarily include sufficient data to reflect the "real" distribution of mortality. Figure 3.1 shows that frequency counts are declining in the older age groups. Thus the "samples" for these age groups are low. This means that statistical tests may consider valid data as outliers. One of these outliers may actually be a "true" longevity record.

Since studbooks contain extensive data on individuals, it is possible to evaluate outliers in terms of data quality. For example, the software program PopLink lists the ten highest "longevity" ages from both living population and death records. Outliers based on artefacts or inaccurate dates of birth and/or death can be excluded from analyses.

3.2.2.2 Data Truncation

Longevity (symbol: Ω) based on a validated outlier may be appropriate "for the record", but it is not recommended to use this value in analyses and/or comparative studies. Simulation experiments based on samples of observed age in marine mammals showed that longevity based on maximum observed age is sensitive to sample size (Barlow and Boveng 1991). The 95th and 99th percentiles[3] were shown to be less sensitive to sample size than longevity based on maximum observed age.

In contrast, sample size is less of an issue in semelparous species with rapid senescent mortality, e.g. post–spawning mortality of Pacific salmon (*Oncorhynchus* spp.) (Morbey et al. 2005), and post–mating male mortality of carnivorous marsupials (Dasyuridae) (Bradley 2003).

Small sample sizes not only affect longevity, but can also affect basic statistics such as mean and median age at death as well. Therefore it is recommended to *truncate* (exclude) the data above the selected percentile and (re-)calculate the quartiles and mean. Table 3.2 presents truncation of the snow leopard data set in the range 1, 5 and 10 percent of the tail of the death–age distribution (*right truncation*). The maximum age in these data sets corresponds to the 99th, 95th and 90th percentiles, respectively.[4]

Since the older ages at death are excluded, values for quantiles, mean and maximum observed age consequently decrease with increasing levels of truncation. The obvious question arises "Which percentage of truncation (or percentile) to

---

[2]A major characteristic of the normal distribution is that the curve is symmetrical around the (arithmetic) mean. Therefore mean, median and mode are all located at the same point.

[3]These percentiles mean that 95 % or 99 % of data ordered by age were included.

[4]The maximum observed ages in truncated data sets can differ from percentiles due to rounding of number of records that are excluded, and the method that is used in calculating the percentiles.

Table 3.2 Summary statistics of death ages (in days), truncated at different percentages, in captive snow leopards from 1890 to 2003

Truncation (%)	1st	Median	Mean	3rd	Maximum	n
None	4	142	1,694	3,372	7,981	1,606
1	4	127	1,637	3,260	7,035	1,590
5	4	99	1,431	2,800	6,185	1,526
10	3	76	1,188	2,244	5,504	1,446

1st first quartile, *3rd* third quartile, *n* sample size

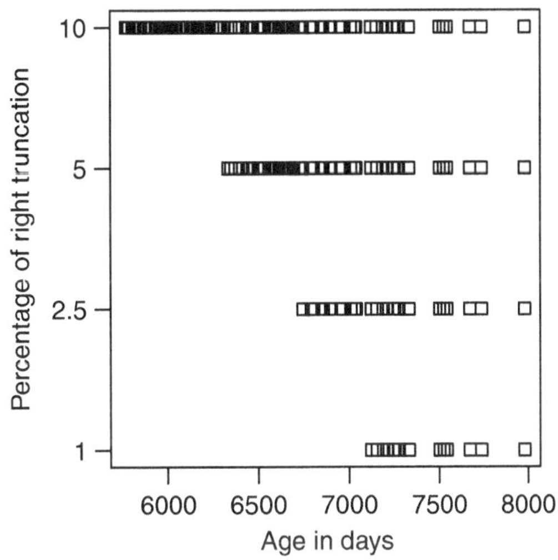

Fig. 3.2 Stripchart of the tail of the mortality distribution for different percentages of truncation in captive snow leopards, from 1890 to 2003

apply?". First of all, it is not the intention to exclude data that would be informative for the distribution of mortality. The answer is to have a closer look at the data in the tail of the distribution (this means the data that would be excluded after truncation).

Figure 3.2 shows one dimensional scatterplots of data in the 1 % (bottom plot), 2.5 % and 5 % (middle) and 10 % (top) of the distribution tail i.e. oldest ages at death. The maximum observed age of 7,891 days could be considered as an outlier (even if the observation is validated). Truncation levels of 5 % and 10 % are clearly too high, as a large amount of data that forms a continuous distribution (regions where strips overlap) and thus likely reflects the mortality distribution, will be excluded. This could more or less have been expected on the basis of the number of 80 and 160 records which need to be excluded, respectively (Table 3.2).

A truncation level of 1 % is still too high for this data set, given that the first block of data is part of the continuous distribution, remembering that the analysis is based on accurate dates. A lower truncation of, for example, 0.5 % would seem to be more appropriate. This example shows that truncation should not be applied a priori without knowledge of the distribution of the data.

3.2.3 Pathologist's View

Table 3.1 shows that *mean age at death* in snow leopards during the period 1890–2003 is 1,694 days ($\approx 4\frac{1}{2}$ years). The mean age at death is also known as *(mean) life expectancy* (Caughley 1977). Life expectancy is often calculated as the arithmetic mean or median age at death. However, calculations based on death records do not necessarily reflect the "true" life expectancy.

The type of data as presented in the previous figures and tables can be compared with the data that veterinary pathologists will collect over years of work. They will have data sets of post–mortem records with information on the age at death. However, pathologists generally do not have data on the living population, such as number of births or number and age of living individuals. Yet, living individuals and individuals that disappeared (*Lost To Follow–up (LTF)*) have ages and, therefore, contribute to life expectancy too! They inform us of having reached a certain age at the date of monitoring or, in the case of LTF, the age at the date when they were last observed.

For example, the mean age of living African and Asian elephants in regional zoo populations was 24.6 and 31 years, respectively. This is higher than calculations of mean age based on dead animals only (16.9 and 25 years) (Wiese and Willis 2004). Combining data sets on ages at death, ages of living individuals at end of study period, and ages of LTF individuals will result in better estimates of life–expectancy (Wiese and Willis 2004).

3.2.4 History and Husbandry

Husbandry in zoological gardens has improved tremendously in the last 100 years, with increased knowledge from practical experience, and from biological and veterinary sciences, including field studies. Living snow leopards most likely "profit" from this improved knowledge of husbandry and have, consequently, a higher life expectancy in recent decades than did those who were imported or were born in the first half of the twentieth century (see Table 3.3).[5]

While longevity records require the full historical data set, it makes more sense to use a time–span which reflects modern knowledge of husbandry when estimating life expectancy. The choice of an appropriate time–span depends on the longevity of the species involved and the year the studbook was initiated. Studbooks on zoo populations often only include historical data on those individuals that are

[5]Although International studbooks for Przewalski's horses (*Equus przewalskii*) and European bison or wisent (*Bison bonasus*) were initiated at the beginning of the 20th century, captive propagation programmes in zoological gardens started on a larger scale in the 1960s. Regional breeding programmes such as the North American SSPs© and European EEPs, which were initiated in the mid 1980s, now include compilation of husbandry manuals.

Table 3.3 Summary statistics of mean age at death in captive snow leopards for different periods of time

Group	1st	Median	Mean	3rd	Maximum	n
1890–1982	4	35	688	610	7,286	396
1983–2003	4	395	2,026	4,260	7,981	1,213

1st first quartile, *3rd* third quartile, *n* sample size

relevant for the pedigree of the population (prior to the date when the studbook was established). This particularly applies to studbooks that were established before computerised in-house registration, linked to the global database at the International Species Information System (ISIS), became active (Flesness and Mace 1988; Flesness et al. 1995). This implies that the data of the "early" years in a studbook may be restricted to those individuals that survived until the age of reproduction and, moreover, have (surviving) descendants in the current population. This leads to the following guidelines in selecting study periods:

1. Include a study period that covers approximately the maximum lifespan (longevity) before the last update of the studbook. For example, the first date for captive snow leopards would be 1981 or 1983, based on a longevity of 20–22 years (see Table 3.1) and the last full update as on 31 December 2003.
2. Start the study period in the year when the studbook was established in the case that the longevity record refers to an earlier date.
3. Start the study period around 1990 for recent studbooks, especially when data have been extracted from the ISIS database (Flesness et al. 1995).
4. Use results of census analyses and census-like analyses such as births and deaths per year (see Chap. 4) to determine the appropriate start and end of the study period. For example, when the number of births outnumber the number of imports (Faust et al. 2003).

Since the International studbook on snow leopards was established in the 1970s, it is possible to analyse a study period that comprises the longevity record: from 1983 to 2003. Summary statistics for lifespan (note: based on death records only) during this period are presented in Table 3.3. These results are compared with results from the entire previous period: 1890–1982. Twenty-five percent of the deaths occur between 0 and 4 days after birth in both periods of time. This indicates that modern husbandry has had no effect on (reducing) perinatal mortality (death around birth) in snow leopards. However, mean life–expectancy has almost tripled from 688 to 2,049 days (Table 3.3).

The data set as used in the above example includes individuals that were born before 1983 (and died before the end of 2003). This means that lifespan as estimated for the period 1983–2003 can have been (partly) determined by husbandry factors in the previous period. The solution to this problem is to include only individuals that were born and died during the study period.

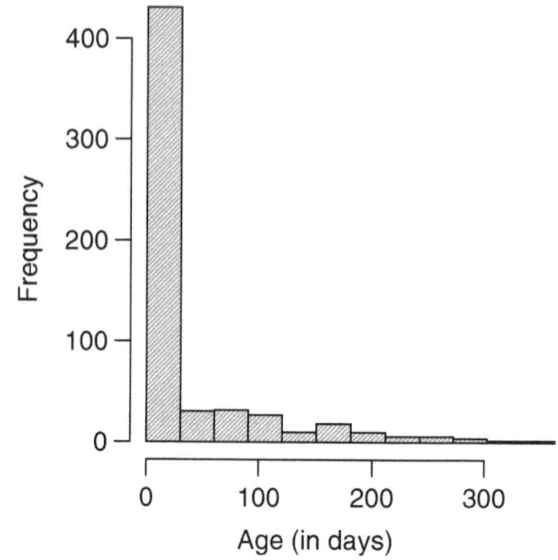

Fig. 3.3 Mortality in the first year in captive snow leopards from 1983 to 2003. Bars correspond with 1 month (30.5 days) intervals

3.2.5 *Early Life Stages*

Causes of death are usually of different origins during different life stages. Figure 3.1 shows that the majority of deaths in snow leopards occurs in the first year. This period covers the early life stages of snow leopards: perinatal mortality, neonatal mortality and juvenile mortality. For example, perinatal mortality, which includes stillbirths and premature deaths, accounts for 25 % (=1st quartile) of deaths in captive snow leopards, while the median age at death, during recent times, is approximately 1 year (see Fig. 3.1 and Table 3.3).

Figure 3.3 presents the frequency counts for mortality in the first year (365 days). This figure shows, as expected from the summary statistics, that the majority of deaths occurred during the first month or neonatal stage. A further refinement, mortality within the first 30 days, is presented in Fig. 3.4a. This shows that the highest mortality within this period occurred in the first week. Thus, a further refinement, mortality within the first week, is recommended.

Figure 3.4b shows that the highest mortality in newborns occurred on the day of birth (which is day 0). Death at date of birth can include stillbirths, spontaneous abortions and death shortly after birth.[6] Life–span analyses within age groups (or life stages) will not reveal the underlying causes of deaths, but can at least provide information to be used in further studies, for example an examination of post–mortem reports. Mortality due to inbreeding depression, however, can be directly detected from studbook data whenever pedigree data are available. This subject will be discussed in Chap. 13.

[6]Premature births are recorded as births in studbooks.

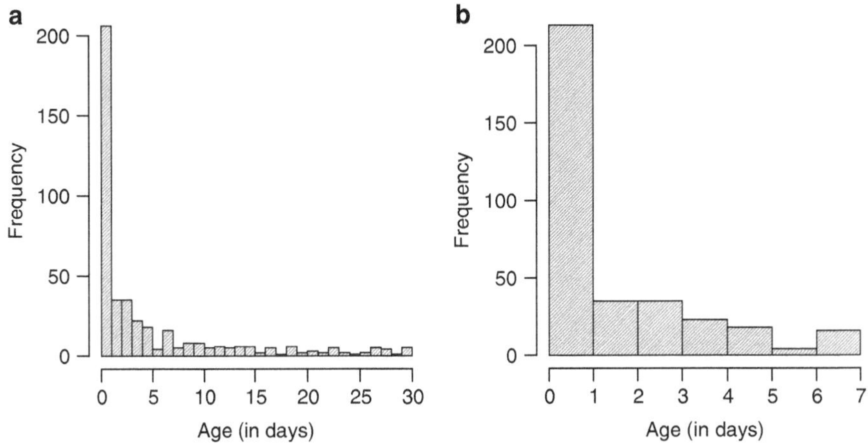

Fig. 3.4 (**a**) Mortality in the first month (30 days) in captive snow leopards from 1983 to 2003. (**b**) Mortality in the first week

The period of weaning can also result in higher mortality. Figure 3.3 shows a slight increase in mortality in snow leopards in the third month, when weaning starts.

3.2.6 *Differences Between Sexes*

The lifespan analyses in the previous sections did not distinguish between sexes. This was to allow the inclusion of individuals of unknown sex. This particularly applies to perinatal mortality, when it may not always be possible to determine the sex through post–mortem examination (for example, due to cannibalism or decomposition). The proportion of individuals of unknown sex in death records is likely to decline, even in species with no obvious sexual dimorphism, during later life stages. Either sex has been determined during life, through observation, examination, karyotyping or applying DNA tests, or through examination after death.

Death during the neonatal and juvenile stage is generally not related to sex, although exceptions will exist. Therefore, analysing lifespan without distinguishing between the sexes seems to be a "safe" approach for the neonatal and juvenile stages. Differences in mortality between sexes are likely to start around the age of sexual maturity, especially in species with sex–biased dispersal. The dispersing sex will, under natural conditions, leave the territory of the parents (or natal group) and potentially be exposed to risks varying from encountering hostile conspecifics and predators, to ending up in unsuitable habitat that increases mortality. The situation in zoological gardens is much more benign than in nature, as such risks are minimal (there is a risk from introductions to conspecifics). Yet, it is the dispersing sex which

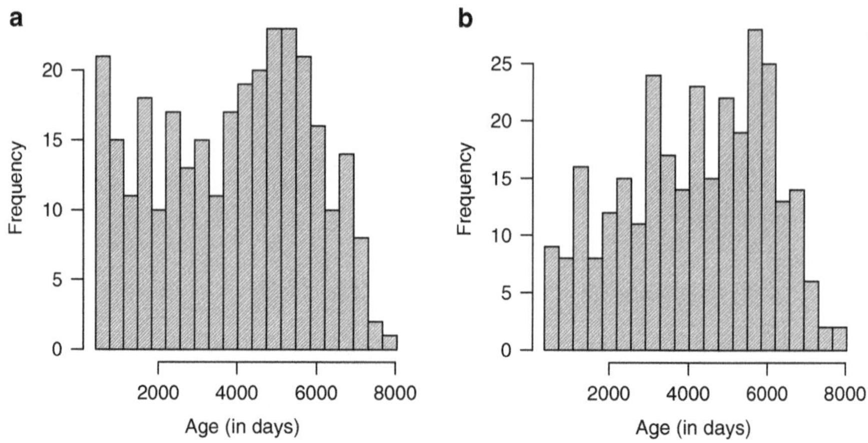

Fig. 3.5 (**a**) Mortality after the first year in captive male snow leopards from 1983 to 2003. Bar width corresponds with 1 year (365.25 days) intervals. (**b**) Females

will eventually be transported to another zoo (or any other ex–situ environment) as part of a breeding programme. However carefully planned and well guided, transport of an animal always entails a mortality risk, whether during transport, quarantine or adaptation to the new zoo environment.

Later differences between sexes can occur during the reproductive ages (see Sect. 3.3). Females can die of complications during pregnancy and delivery, or during egg-laying. Both males and females, depending on the social system, can be involved in dominance fights that can result in fatal injuries.

The proportion of neonatal and juvenile mortality can make it difficult to visualise mortality in later stages (see Fig. 3.1). Therefore, it is recommended to exclude the data of these stage groups from the data set. Figures 3.5a and 3.5b show mortality in male and female snow leopards which have survived the first year, respectively.

The frequency count of male deaths in the first year is higher than in females, resulting in a different scale on the frequency axis. This makes it difficult to interpret whether male and female mortality differ over the entire lifespan.

The Quantile-Quantile (*Q–Q*) plot is a graphical method that can be used to visualise potential differences between two distributions (R Core Team 2015). Figure 3.6 shows a Q–Q plot for mortality in male and female snow leopards. When the quantile points are positioned approximately on the 45–degree reference line (the dashed line), the distributions of mortality for males and females are the same. However, statistical tests are required to determine whether deviations from this central reference line imply significant differences.

Lifespan data do not follow a normal distribution (as discussed in Sect. 3.2.2). Therefore parametric tests, such as the Student's t test, cannot be applied to test for differences in mean age at death between males and females. A nonparametric test, such as the *Wilcoxon two–sample test* (Sokal and Rohlf 2012), can be applied to test

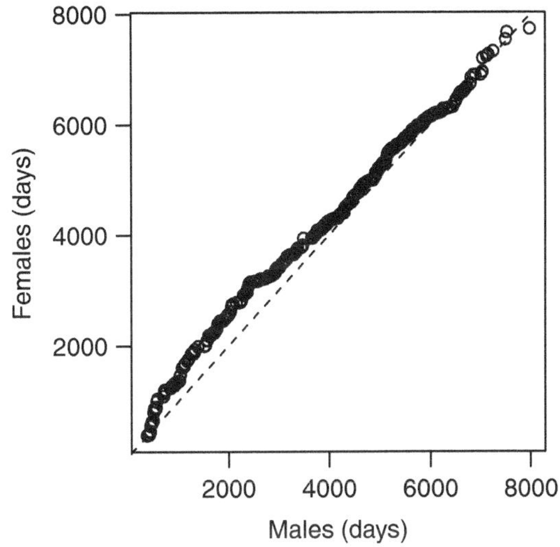

Fig. 3.6 Q–Q plot of male and female mortality after the first year in captive snow leopards from 1983 to 2003

for differences between males and females. The null hypothesis (H_0) is that both samples are drawn from the same (statistical) population.

The Wilcoxon two–sample test on male and female mortality in snow leopards ($W = 42,231.5$, $p > 0.05$) does *not* reject the null hypothesis.[7] This means that no difference, between male and female mortality in snow leopards which survived the first year, could be detected with this test.

It is recommended to present mortality data per age group proportional to the total number of deaths, whenever comparisons between distributions need to be made. The proportion of total number of deaths in age group x refers to the probability (chance) that a new–born dies in that age group. This parameter is known as age–specific *mortality* (symbol: d_x) in life–tables (see Chap. 7).

Lifespan analyses as presented in this section are based on frequency counts of age at death. The number of individuals that were actually born and survived are ignored (see Sect. 3.2.3). Chapters 7 and 10 will present methods that include both death records and data that include number of births and survival per age group.

Nonparametric tests are not free from the basic statistical requirement of independent sampling. The litter effect, maternal effect and individuals sharing conditions that are specific for individual zoos (e.g. husbandry practices), are examples of *pseudoreplication* that can affect the validity of statistical results (see Hurlbert 1984; Lombardi and Hurlbert 1996).

A list with ages at death does not suffice to determine whether pseudoreplication occurred. Additional information on individuals, such as location of birth and lifespan of litter mates, would be required. The life–table methods that are applied to studbooks use such information, and, therefore, are more reliable for comparing

[7]The statistical terminology is to *reject* and to *not reject* the null hypothesis.

mortality differences between groups (such as males and females) than the "crude" lifespan analyses are. Nevertheless, the "crude" lifespan analyses provide a quick way to assess longevity and to gain an initial overview of age–related deaths in the population.

3.3 Reproductive Lifespan

Reproductive lifespan refers to the age of individuals at time of birth (or hatching) of their first and last offspring. The PML software (Princée 2014), which is used in this section, includes all reproductive ages. This means that additional parameters, such as median and mean age at reproduction, can be estimated too.

Since differences between sexes (can) exist, this analysis is usually carried out separately for males and females. The major use of this analysis is to describe the biological age limitations of reproduction. Therefore, offspring of both sexes are included in the analyses. This method differs from age–specific fecundity rates, which are generally based on offspring of the same sex as the parent (see Chap. 7).

Reproductive lifespan is calculated from birth data of individual offspring and not from litter dates. This means that potential effects of the parent's age on litter size is reflected in the results.

Figures 3.7a and 3.7b present histograms of reproductive ages in captive male and female snow leopards during the period 1983–2003, respectively. These figures shows that reproduction increases rapidly with age to a peak and then drops

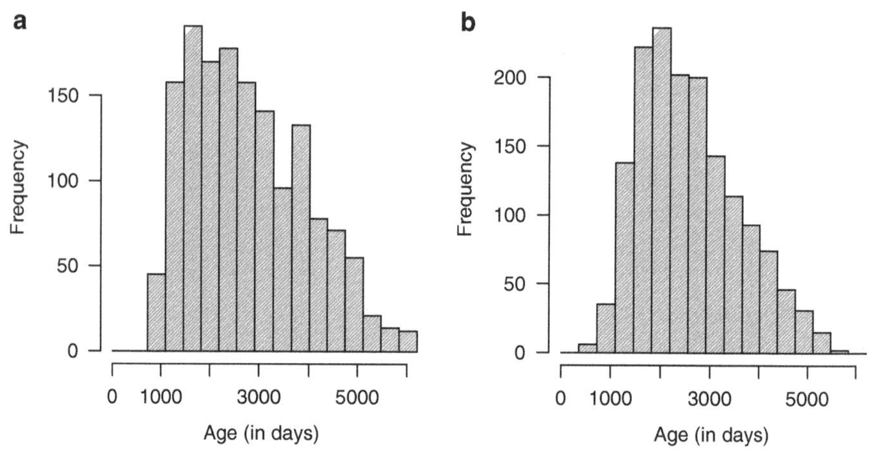

Fig. 3.7 (**a**) Reproductive lifespan in captive male snow leopards from 1983 to 2003. Frequency counts refer to numbers of births to males per age group. Bars correspond to 1 year intervals. (**b**) Females

Table 3.4 Summary statistics of age at first reproduction and reproductive life-span (in days) in captive snow leopards from 1983 to 2003

Analysis	Group	Minimum	1st	Median	Mean	3rd	Maximum	n
Lifespan	Female	500	1,824	2,527	2,603	3,272	5,573	1,557[a]
	Male	733	1,819	2,569	2,779	3,656	6,180	1,521[a]
First age	Female	500	1,449	1,819	1,953	2,206	5,444	281[b]
	Male	733	1,436	1,830	2,103	2,627	5,506	262[b]

1st first quartile, *3rd* third quartile, *n* sample size
[a] Number of births
[b] Number of breeders

gradually. Summary statistics, presented in Table 3.4, provide a way of expressing the characteristics of reproductive lifespan in numbers. Note that both the minimum and maximum ages can be outliers (see discussion in Sect. 3.2.2).

The first quartile and median values are quite similar for both males and females (Table 3.4). However, the mean and third quartile show that reproduction in females tends to decline more rapidly than in males (see also Figs. 3.7a and 3.7b). The Wilcoxon two–sample test shows that male and female reproductive lifespans differ significantly ($W = 1,266,411, p < 0.001$).

The arithmetic mean age at reproduction is also known as *generation time* (symbol: $\bar{T}$). This value is used in population genetics to determine the expected loss in genetic variation – which is generally expressed as loss per generation – over a given period of time (see Chap. 14). Note that generation time is generally calculated from life–table analyses (Chap. 7).

It is important to note that the method presented here is based on individuals that have produced offspring. This type of analysis does not provide information on the actual fraction of individuals within age groups that have reproduced. To do this requires calculation of age–specific fecundity rates, as will be discussed in Chap. 7.

3.3.1 Age at First Reproduction

The minimum observed age at reproduction can be an outlier. Therefore, it is not recommended to use this value as age at first reproduction (symbol: T_1). The arithmetic mean or median age at first reproduction should be used instead (symbols: $\bar{T}_1$ and $\tilde{T}_1$, respectively). Figure 3.8 presents age at first reproduction in male snow leopards. It shows that the majority of males do not start breeding until the age of 3–5 years.

Table 3.4 presents summary statistics for age at first reproduction in both male and female snow leopards. The median age at first reproduction for both sexes is 5 years.

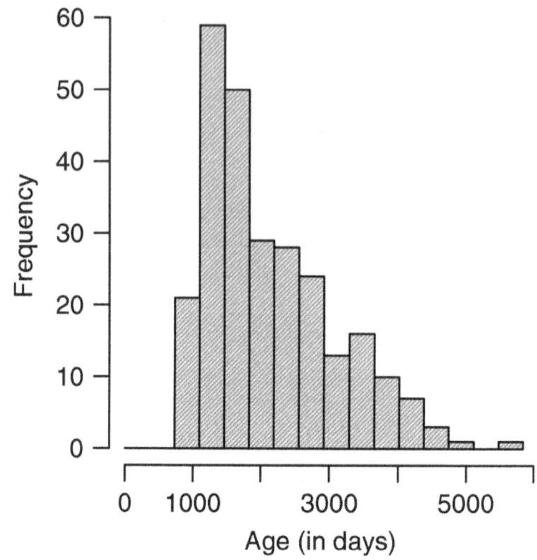

Fig. 3.8 Age at first reproduction in male snow leopards from 1983 to 2003. Bar width corresponds to 1 year intervals

3.3.2 Artefacts

Reproduction in captive populations is influenced by management. For example, logistics such as the time to obtain permits for international transports (for example CITES permits) and/or length of quarantine, can delay the physical possibility for reproducing. This means that the biological first age at reproduction can be lower than observed in the studbook data. Likewise, artificial reproductive techniques can prolong reproductive lifespan, even beyond the death of individuals. This refers to the concept of the "frozen zoo" when sperm and/or embryos of individuals which have died are used in artificial reproduction (Benirschke 1984; Clarke 2009). This means that data referring to offspring from such techniques need to be excluded from analyses.

The main aim of analysing ages at reproduction is to provide an impression of ranges in reproductive lifespan and to indicate potential optimal ages (as indicated by the mean age) for reproduction. This information is important for selecting age groups for reintroduction/translocation.

3.4 Litter Size

Litter size in the context of studbooks generally refers to the number of offspring born to a female of a live-bearing species or to the number of hatchlings from a single clutch of an egg-laying species.[8] Although zoological gardens maintain

---

[8]Studies on wildlife populations usually consider clutch size rather than hatchlings.

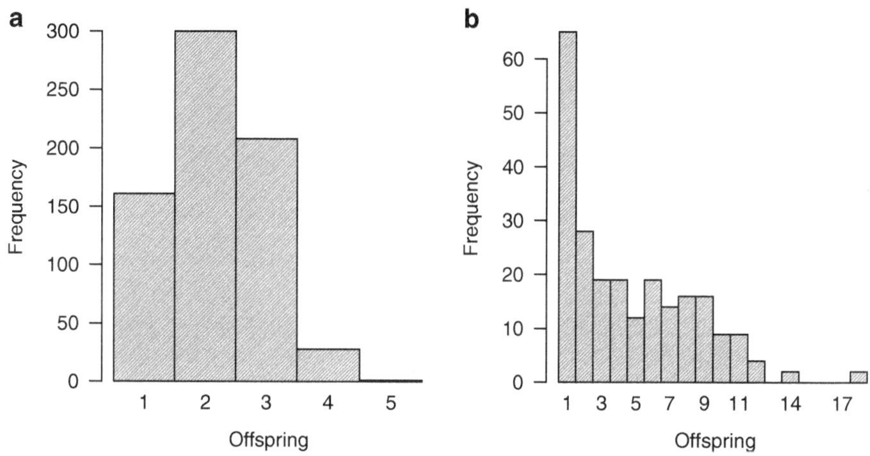

Fig. 3.9 (**a**) Litter size in snow leopards from 1983 to 2003 (n = 700 litters). (**b**) Litter size in African wild dogs from 1986 to 2006 (n = 234 litters)

records on clutch size (and incubation time), these data are not included in studbooks (Thompson et al. 1997). The life of an individual in a studbook starts at the moment of birth or hatching. Stillbirths, spontaneous abortions and hatchlings which died during hatching are included. However note that these data only refer to individuals that have actually been observed (or their remains detected).

Figure 3.9a shows the frequency distribution of litters in captive snow leopards in the period 1983–2003. The median and mean values are 2.000 and 2.146, respectively (n = 700). Litters of four offspring are rare and those of five offspring seem to be exceptional.

Histograms of frequency counts provide an intuitive way to judge the significance of a distribution pattern. Stochastic population modelling programs, such as VORTEX (Lacy et al. 2009), require that litter sizes are provided as fractions or percentages which are used as probabilities. In this, be aware that data presented as fractions may not be that informative whenever based on small sample sizes.

3.4.1 Assessing Litter Size

Litter sizes in the EEP population of African wild dogs show a wide range from 1 to 18 offspring (Fig. 3.9b). The median and mean values are 4 and 4.7, respectively (Table 3.5). However, the first quartile – accounting for 25 % of the litters – is only one offspring per litter. One could take these results at face value, but do they actually reflect the reproductive biology of African wild dogs?

African wild dogs deliver in a den and the litter size generally remains unknown until the puppies emerge after 3–4 weeks. This situation is not much different in

Table 3.5 Summary statistics of litter size based on record dates of births and surviving puppies at 21 days (when they are assumed to emerge from the den) in the EEP population of African wild dogs, from 1986 to 2006 ($n = 234$)

Group	*1st*[a]	Median	Mean	*3rd*[a]	Maximum
Birth	1	4	4.7	7	18
Den emerge[a]	1	1	3.4	5	14

1st first quartile, *3rd* third quartile
[a] Emerging from den is around 21 days (Estes 2012)

zoological gardens, where keepers tend to leave the den/box undisturbed during the breeding season. This means that only puppies which survived these first weeks are observed and, thus, counted. By the time the den can be inspected, remains of puppies which have died will not be found. Thus, the actual litter size is likely to be underestimated in species which deliver in dens that are not accessible for observations.

Infra–red cameras installed in dens can be used to determine litter sizes more accurately. This technique is used in zoological gardens, for example to record birth in polar bears (*Ursus maritimus*) and wolverines. Recording vocalisations of puppies has proved to be a good technique to determine litter size in African wild dogs at an early stage. This information will contribute to better biological knowledge of ranges in litter sizes and perinatal mortality.

The techniques as described above may, depending on costs involved, only be applied to a small number of dens (and/or litters). This means that in these cases, accurate data can actually create bias in estimating the overall, observed, distribution (and mean, median etc.), of litter sizes. This bias is determined by mortality in the den, accounting for potential differences between actual litter size and litter size as observed when offspring emerge from the den. This bias will affect the biological interpretation of litter sizes, and, therefore, comparison with data-sets that are solely based on observation of surviving offspring.

A solution to cope with "missed" offspring is to calculate litter sizes as number of surviving offspring around the date that they are expected to emerge from the den i.e. ignore the problem. This method is applied in collecting data on African wild dogs in the wild (e.g. McNutt and Silk 2008). Puppies of the African wild dog emerge from the den around 3 weeks (Estes 2012). Litter size, based on surviving puppies in the EEP population at the age of 3 weeks, is presented in Table 3.5.

3.4.2 Date Range in Litter/Clutch Size

Calculating litter sizes from studbook data is a retrospective method in which offspring born to a female within a biologically meaningful date range are counted as belonging to the same litter. The method is generally straightforward in assessing litter sizes in live-bearers, as births of the first and last offspring are likely to occur within 24–48 hours. It becomes more complex, however, in egg–laying species when eggs are laid over several days, or new eggs are laid to replace lost ones. This may include management techniques to increase reproduction by removal of eggs which are incubated artificially or by foster parents. This means that eggs from a single clutch may hatch at different times.

Artificial incubation can also create wider ranges in hatch (birth) dates within a clutch as, for example, a group of eggs could have been divided over different incubators to experiment with humidity and temperature. The problem as described above can be solved for species with distinct birth seasons (birth–pulse model). The range in hatch date within a single clutch could in those cases simply refer to the length of the birth season.

Analyses of seasonality in births (hatching) and inter–birth (hatch) intervals can be helpful in determining reasonable ranges in dates which refer to the same litter or clutch (see Sects. 3.5.2 and 3.6).

3.5 Seasonality

Seasonality refers to events, such as migration and breeding season, which occur in a regular pattern within a year. These seasons could be spring, summer, autumn and winter in the temperate zones or rainy and dry seasons in tropical zones. Studbooks contain data on two potentially seasonal events: births and deaths. The seasonality in these events, and methods to test their (statistical) significance, will be discussed in the following sections.

3.5.1 Mean Date

Days, weeks, months and quarters can, conveniently, be numbered from 1 to 365,[9] 1 to 52, 1 to 12 and 1 to 4, respectively. It may seem that it would be easy to calculate the mean season and its standard error. However, seasons are annually repeating events which do not necessarily start or end within a calendar year (for example, winter in the northern hemisphere). Therefore, re–coding of dates to a scheme where the first date has a lower "value" than the last date is required in order to calculate the mean date and its variance (see e.g. Caughley 1977). This is also the underlying method of *circular statistics* (Pewsey et al. 2013; Zar 1984). Time intervals, whether days or quarters, are re–scaled to 360 degrees on the circle. This statistical method is presented in more detail in Chap. 19. The R package *circular* (Agostinelli and Lund 2013) is used for seasonality analysis in this book.

[9] 366 days if leap years are taken into account.

3.5.2 Seasonality in Births

The reproductive season encompasses the period from mating to birth or hatching. Knowledge of seasonality in breeding, especially in mating activities, is important for management of captive populations.[10] Cooperative breeding programmes involve exchanges of animals between zoos in order to minimise inbreeding and genetic loss (Lacy et al. 1995). Appropriate timing in planning these exchanges is important, especially for species with a short reproductive lifespan (Sect. 3.3). One could easily miss a mating season.

Two major breeding systems are generally described (Caughley 1977):

1. Birth–flow: the rate of breeding is constant throughout the year.
2. Birth–pulse: all offspring are produced on 1 day, the date being the same from year to year.

In reality, these systems do not necessarily fully apply to species or populations. However, the reproductive season may come close to one of them. Knowledge of the breeding model that "fits" best with a species is relevant to determining census dates, and will be discussed in Chaps. 4 and 7.

Information on seasonality in mating season may reveal the underlying mechanism(s) which potentially triggers reproductive behaviour, as will be illustrated in this section. Such information is important as captive conditions (climate, day-light regime) often differ from the natural habitat. Although modern zoos will register mating activities of species in the in–house registration systems, these data are not necessarily included in studbooks.

Studbooks generally only register "hard" observations such as date of birth. This limits the use of studbook data for detecting mating season. Therefore, the mating season needs to be deduced, using existing natural history data on gestation or incubation length, counting backwards from the date of birth.

Seasonality in births in species that have singletons is based on individual dates of birth. In species with multiple births, only a single date, e.g. first or median, from a litter is used in order to avoid pseudoreplication (Hurlbert 1984).

Species that predominantly produce singletons can still produce the odd twin or triplet. For example, 6 out of 793 individuals in the blesbok population sample are from twin litters. Including the "extra" half of these twins as individual records is not expected to change the results, in this particular case.

Figure 3.10 shows a circular graph of date of births (litters) in the European zoo population of the wolverine. This figure shows a rather distinct season from 1 February until 8 March, with a peak in births in early March. The Rayleigh test shows a significant mean date of birth at day 51, which is 21 February ($r = 0.99$, $p < 0.001$, $n = 42$).

[10]The breeding season of species with delayed implantation is actually divided in a mating, an "implantation" and a birth season, sometimes separated by months.

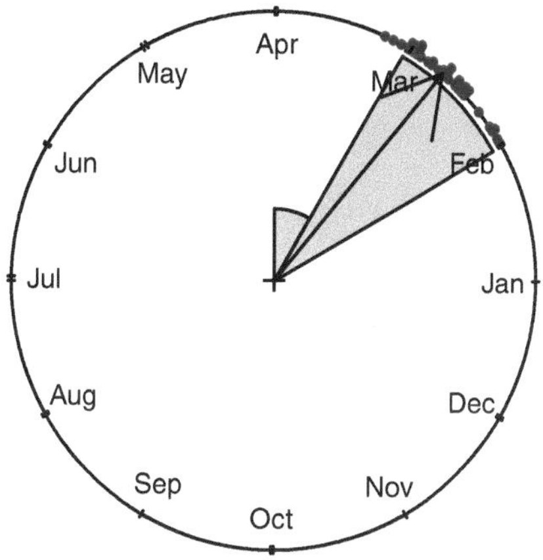

Fig. 3.10 Season of birth based on litter dates in the captive population of wolverines in European zoos from 1963 to 2001. *Stacked points* refer to births per day of year; the area sectors in the rose diagram represent relative frequencies of births per month; and the *arrow* indicates the mean date of birth (21 February)

3.5.2.1 Sex at Birth

Data on births can include a relatively large proportion of animals of unknown sex (see Sect. 3.2). This refers especially to those species where post–mortem examinations of neonates is not feasible, for example in the case where the remains are disposed of by the parents. Excluding animals of unknown gender may reduce the sample size required to detect seasonality – remember the number of records in studbooks on endangered species is relatively low. Therefore, it is recommended to include all sexes in the initial analyses of seasonality in births.

Separate analyses of seasonality per sex may not be relevant for species with *genotypic (chromosome–based) sex determination (GSD)*. These are typically warm–blooded (or homeothermic) species, but do not necessarily exclude cold–blooded (or poikilothermic) species. However, cold–blooded species can be subject to embryonic environmental conditions which determine the sex of offspring. For example, *temperature–dependent sex determination (TSD)* has been observed in various fish and reptile species (Valenzuela 2004, 2008).

Differences in temperatures do not only determine sex, but can affect later reproductive success of the hatchlings as well. For example, experiments in the Jacky dragon (*Amphibolurus muricatus*), that involved decoupling of sex and incubation temperature by hormonal manipulations, resulted in males and females that developed outside the "sex–specific" temperature range. The study showed that only females and males that were born at natural (sex-specific) temperatures reproduced successfully (Warner and Shine 2008).

Captive programmes which maintain poikilothermic species in outdoor breeding facilities (for example in range countries), should certainly take differences in seasonality between sexes into account.

Seasonal trends in sex–ratio have been observed in birds, especially in raptors, e.g. nests produced early in the season in the European kestrel (*Falco tinnunculus*) have excess of males; late nests have excess of females (Dijkstra et al. 1990). Male calves in reindeer (*Rangifer tarandus*) are heavier in weight than females,

Table 3.6 Seasonality in births of female and male blesbok in the northern hemisphere from 1970 to 2008

Group	2.5 %	Mean	97.5 %	n	p
Female	14 June	19 June	24 June	556	<0.001
Male	21 June	25 June	30 June	598	<0.001

2.5 %, 97.5 % confidence limits, *n* sample size, *p* probability of Rayleigh statistic

and gestation length is longer for males than for females (Mysterud et al. 2009; Weladji and Holand 2003). This could suggest potential differences in mean birth dates between sexes.

Table 3.6 shows that the mean dates of birth of female and male blesbok in the northern hemisphere are 19 and 25 June, respectively. The birth season in both sexes is not uniform (Rayleigh test, $p < 0.001$). However the 95 % confidence intervals overlap the period 21–24 June, which may indicate that these mean values are not significantly different. The "Watson's two sample test" (Zar 1984) confirms this suspicion: the null hypothesis "the seasons are the same" is not rejected ($p > 0.10$).

The subject of sex bias in season of birth is related to the complex subject of sex allocation, in which social rank and physical condition of females also play a role (see e.g. Cockburn et al. 2002; Komdeur 2012). Tests on differences in season of birth between sexes in species with multiple births can be achieved by weighting litter dates for number of males and females.

3.5.2.2 Premature Births

Studbooks do not always register whether stillborns or individuals that die shortly after birth refer to premature births. Refinement of seasonality analyses by including data on viability of offspring could help to detect premature births. For example, a study on birth timing in Californian sea lions in captive facilities in North America showed that mean dates of birth in stillborn and non–viable (lived less than 1 day) offspring occurred 4 weeks before the mean birth date (13 June) of viable (lived at least 1 day) offspring (Temte 1993). In other words, the stillborn and non–viable births are likely to refer to premature births.

Similar analyses, but using circular statistics, have been carried out on the European studbook population of Californian sea lions for this book. Individuals that were born alive but died the same day can not be distinguished from stillborn in basic studbook data sets. These individuals are grouped in this study under the category "day–zero–death". The survival criterion in the Temte's (1993) viable category is extended from at least 1 day to at least 6 months ("weaning age"). Offspring that survived 1 day but died before being weaned are grouped in the category "non–weaned".

Table 3.7 presents results of seasonality analyses on these categories. The mean birth date of day–zero–deaths is 16 and 22 days earlier than those that died before and after reaching the weaning age, respectively. Offspring that died before the weaning age were born on average 6 days earlier. The non–parametric *pairwise*

Table 3.7 Seasonality and viability in births in captive Californian sea lions in Europe from 1978 to 2008

Group	*n*	Day	Date	p
Weaning–age	695	161	10 June	< 0.001
Non–weaned	96	155	4 June	< 0.001
Day–zero–death	138	139	19 May	< 0.001

n sample size, *day* day of year, *p* probability of Rayleigh statistic

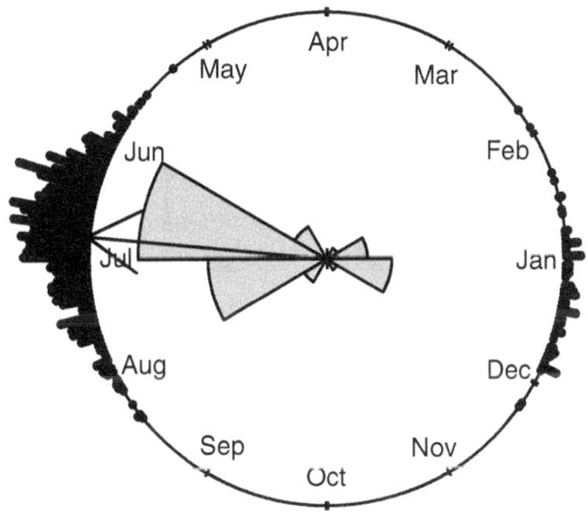

Fig. 3.11 Season of birth (litters) in the global captive population of Nepalese red pandas from 1986 to 2006. *Stacked points* refer to births per day of year; the area sectors in the rose diagram represents relative frequencies of births per month; and the *arrow* indicates the mean date of birth (26 June)

Wilcoxon rank test with Holm's correction for multiple comparisons (R Core Team 2015) was applied to these data. Season of birth in day–zero–death category differs significantly from non–weaned and weaned–age categories (both $p < 0.0001$). Seasonality in the non–weaned category also differs significantly from the weaned category ($p = 0.0002$).

It is beyond the scope of this book to discuss the possible reasons for differences in results of these studies. The main importance of this type of information is that analyses of seasonality in birth can contribute to knowledge of natural history – information which can be used in daily management. For example, zoo keepers may find remains of a new–born, for which it is not clear whether it was a premature birth, stillbirth or whether it was killed by conspecifics. The last cause may imply that a study on potential social problems is required. However, knowing that the date of birth refers to a premature birth changes these perspectives.

3.5.2.3 Latitude

Figure 3.11 shows seasonality of birth (litters) in the international studbook population of Nepalese red pandas from 1986 to 2006. A mean birth season in the period June–July and a second, less pronounced, season in December–January seems to occur. Circular statistics, however, will not detect that second season, as the number of litters in the "first" season outnumbers the "second" season. The

Table 3.8 Results of the Rayleigh test on seasonality in births (litters) in captive Nepalese red pandas from 1986 to 2006

Regions	n	$\bar{a}$	Day	r	p
All regions	787	3.05	177 (26 Jun)	0.764	< 0.001
Northern hemisphere	644	3.04	177 (26 Jun)	0.972	< 0.001
Southern hemisphere	79	−0.15	358 (23 Dec)	0.975	< 0.001

n sample size, $\bar{a}$ mean angle in radians, *r* resultant vector, *p* probability of Rayleigh statistic

Rayleigh test shows a significant ($r = 0.764$, $p < 0.001$, $n = 787$) mean season (for all regions) corresponding with day 177 which is 26 June (see Table 3.8). This example emphasises the importance of visualising data, such as through circular plots, to detect bi–modality.

However, it would be a mistake not to study the "second" season in more detail. International studbooks include individuals in zoos in different parts of the world, with different seasonal characteristics. For example, it is summer in Australia at the time it is winter in Europe. Table 3.8 shows significant differences between breeding seasons of red pandas in the Northern and Southern hemispheres with mean dates of birth (litter) of 26 June and 23 December, respectively. This explains the bi–modality in birth season.

3.5.2.4 Photoperiod

The day number as used in the previous examples refers to the Gregorian calendar (i.e. day 1 = 1 January, day 365/366 = 31 December[11]). This definition is useful, though most would prefer to use the date rather than the day number, in terms of management of populations. However, when studying the mechanisms behind seasonality in birth, it may be better to use day numbers that refer to the astronomical calendar which starts at 20 or 21 December (shortest and longest day, in the Northern and Southern hemispheres, respectively). The mean date of birth in Nepalese red pandas is, according to these data, 4–5 days after the longest day (June and December solstices, in the Northern and Southern hemisphere, respectively).

Geographic differences in seasonality in births can be more subtle than observed in Nepalese red pandas between the Northern and Southern hemisphere (Table 3.8). A smooth change in birth (pupping) date across latitude was observed in captive Californian sea lions in USA facilities – although the origin of most founders could be traced to the same wild population (Temte 1993). This study showed a strong correlation between photoperiod (day length) triggering birth (birth timing) in this species across the different latitudes in North America.

[11] This depends on leap years.

3.5.3 Seasonality in Deaths

Under natural conditions, seasonality in death is expected to be as common as seasonality in births. Whether a climate with dry or rainy seasons, or a temperate climate with four seasons, they both have periods with shortage of food and/or water during which mortality increases. Captive populations, especially those which are maintained in zoological gardens, are not exposed to the harsh conditions of dry or winter seasons or migration. Does this mean that it is not necessary to analyse seasonality in deaths? The straight answer is no!

Captive populations which are located outside the natural habitat are exposed to seasons which differ from their natural environment. Seasonal effects may be subtle and are not necessarily detected. For example, a single mortality case in a waterfowl species in late summer at a single zoo may not trigger extra attention. However, seasonality analysis of studbook data may show that deaths in late summer are more common than expected. Such a result could initiate a follow–up study looking into post–mortem reports.

Figure 3.12a shows seasonality in deaths in the EEP population of African wild dogs. This analysis includes all recorded deaths from 1986 to 2006. Although deaths are more distributed over the year, the period November–January shows a concentration, close to the observed season of births. The Rayleigh test rejects a uniform distribution, with a mean day of 2 January ($p < 0.001$, $n = 447$).

Neonatal mortality contributes substantially to the total number of deaths in a population (see Sect. 3.2). Consequently, seasonality in deaths will occur whenever seasonality in births exists. Furthermore, neonatal and juvenile deaths in species with litters can be subject to the litter effect, i.e. litter mates have a higher or lower chance of dying than would be expected from the mortality rate (see Chap. 19 for more details).

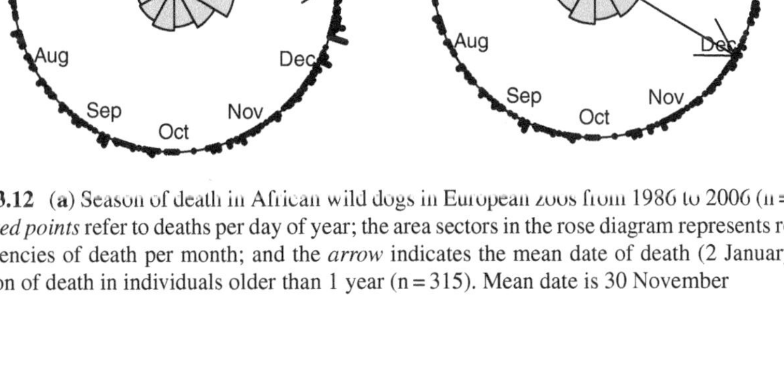

Fig. 3.12 (**a**) Season of death in African wild dogs in European zoos from 1986 to 2006 (n = 447). *Stacked points* refer to deaths per day of year; the area sectors in the rose diagram represents relative frequencies of death per month; and the *arrow* indicates the mean date of death (2 January). (**b**) Season of death in individuals older than 1 year (n = 315). Mean date is 30 November

Deaths in this population are distributed more uniformly over the year when those that occurred in the first year (neonates and juveniles) are excluded (see Fig. 3.12b). However, a seasonality with a mean date of 30 November ($p < 0.001$, $n = 315$) still exists.

3.6 Inter–Birth Interval

Inter–birth interval is the time–span between successive litters (or clutches) produced by the same female. Various biological factors such as birth–flow or birth–pulse breeding systems (see Sect. 3.5), gestation or incubation length, weaning period and infant survival can influence inter–birth interval. The effect of environmental factors such as food availability, especially in combination with density–dependent reproduction, may seem to be restricted to wild populations. However, mark the words "seem to be", as the zoo environment clearly can also have an impact on inter–birth intervals.

Figure 3.13 shows the distribution of inter–birth (litter) intervals, grouped in classes of 1 month, in captive snow leopards. It shows a sequence of more or less equally distributed peaks. Snow leopards – in zoos in the Northern hemisphere – have a significant birth season in May (Rayleigh test: $r = 0.844$, $p < 0.0001$, $n = 988$). This means that the observed peaks in inter–birth intervals refer to sequences of these annual birth seasons. The median and mean inter–birth intervals are 690 and 712 days, respectively (Table 3.9). This means that 50 % of the litters are produced (nearly) biennially.

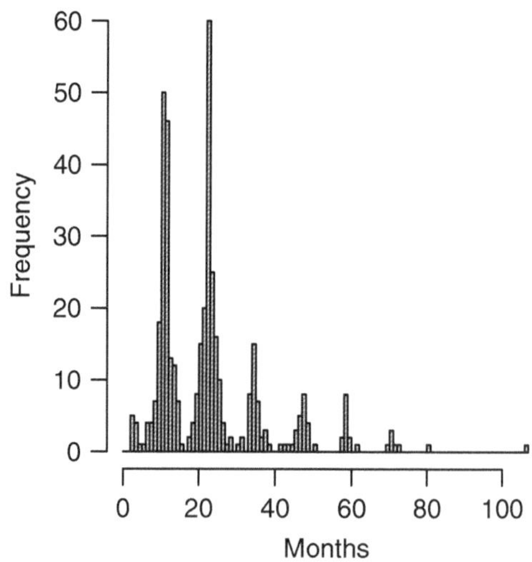

Fig. 3.13 Inter–birth interval in captive snow leopards from 1983 to 2003

3.6.1 Management Effects

The minimum observed inter–birth interval in the snow leopard studbook is 114 days (Table 3.9), which is close to the gestation length of 96 days (Blomqvist 2004). May is the main season of birth, but some captive snow leopards have produced litters in August and September. Given that this species originates from the Himalayas, one would not expect new litters during late summer in the wild. Moreover, weaning age in snow leopards is around 3 months (Blomqvist 2004). However, it is common knowledge in zoological gardens that "strict" seasonal breeders may produce a second litter when the previous litter is not viable or when the newborns have been separated from the mother (and foster–reared or hand–reared). This illustrates the impact of the zoo environment on inter–birth intervals.

Cooperative breeding programmes involve transfers of animals between zoo locations in order to avoid inbreeding and to optimise maintenance of genetic variation (Lacy et al. 1995). Inter–birth interval can be affected when females are transferred to another location between mating seasons. Individual adaptation to the new environment and/or the species' specific biology, for example adaptation to photoperiod (see Sect. 3.5.2), can cause a temporary increase in inter–birth interval.

A more "natural" distribution of inter–birth intervals can be obtained by excluding intervals when (1) offspring of the previous litter are hand–reared or foster–reared; (2) all offspring of the previous litter died before weaning; and (3) the previous litter was delivered at a different location.

Table 3.9 shows the results of analysis of "natural" inter–birth intervals in captive snow leopards. These data show a shift towards longer inter–birth intervals, compared to the data set that includes all litters. The minimum observed interval of 125 days, however, still refers to a second litter within the same year. This indicates that a female became pregnant 1 month after giving birth, but before weaning her (viable) offspring. It is important to evaluate the data to determine whether this outlier is an exception, an artefact such as kittens that have been taken from the mother after a month, or that the date of birth or type of rearing (of the previous litter) is a typing error. In this, analyses of natural history elements act as feedback mechanisms to improve data quality of studbooks.

Table 3.9 Summary statistics of inter–birth interval in snow leopards from 1983 to 2003

Rearing	n	Minimum	25 %	Median	Mean	75 %	Maximum
All	425	114	373	690	711.6	783	3,287
"Natural"[a]	217	125	627	729	842.1	1,087	2,527

25 % first quartile, *75 %* third quartile

[a] Only intervals between litters that match criteria to minimise effects of the captive environment and/or management are included (see text)

Table 3.10 Summary statistics of "natural" inter–birth interval in snow leopards during different periods of time

Period[a]	n	Minimum	25 %	Median	Mean	75 %	Maximum
1974–1983	40	330	363.8	376.5	529	703	1083
1994–2003	78	173	693.2	771.5	912	1104	2208

n sample size, *25 %* first quartile, *75 %* third quartile
[a] Refers to 1 January of the first year to 31 December of the last year

The maximum observed inter–birth intervals in both "all" and "natural" data sets (Table 3.9) are detected as outliers in statistical tests (see Komsta 2011). These outliers are likely to be – within the biological context – artefacts due to captive management. For example, cooperative breeding (management) programmes identify genetically important individuals that need to reproduce. This can refer to older females that have not been in a breeding (mating) position for some time since their previous litter.

Management programmes that involve birth control, either by separating sexes and/or by applying chemical contraception, will result in prolonged inter–birth intervals. Although these type of data are registered at the individual zoo level, they are not widely included in studbooks. But even when studbooks include "birth control" data it is likely that early data (prior to computerised registration) are not included. That limits the option of refining the data set in order to analyse "natural" inter–birth intervals.

However, this is not the end of the story! The median and (arithmetic) mean values for inter–birth intervals are expected to increase during periods of time when birth control is implemented on a wider scale. Census data, which will be discussed in Chap. 4, can be used to indicate when birth control has likely been implemented. One could also apply a "rule of thumb" by assuming that birth control was not actively implemented before cooperative breeding programmes were established (in the mid 1980s). In other words, one would expect lower values for the median and the mean in early studbook data.

Table 3.10 shows results of analyses on the snow leopard studbook for different (10 year) periods (1974–1983 and 1994–2003). The criteria for "natural" data sets have been applied in both analyses. Median and mean values for inter–birth intervals in the period 1994–2003 have almost doubled compared to the period 1974–1983. The nonparametric Wilcoxon two–sample test, as discussed in Sect. 3.2.6, can be applied to test whether these differences are significant. The (one–sided) test shows that inter–birth interval in the period 1974–1983 is significantly shorter ($W = 604.5$, $p < 0.001$) than in the period 1994–2003. This is an indication that behaviour and/or population management has increased inter–birth interval in snow leopards.

3.6.2 Egg–Laying Species

Section 3.4.1 discussed the problems that arise in calculating clutch sizes in egg–laying species, where eggs may be laid, either naturally or through manipulation, over a period of time. The same problem occurs in calculating inter–birth intervals. In this, the main question is "What is a clutch?".

Analyses of seasonality in hatching, whether based on individual hatch dates or clutch dates, will show if a species follows a birth–pulse model (distinct breeding season). The season may be stretched because of the longer period of egg–laying, but one would expect to find a significant cluster of hatch dates. A simple approach would be to count all offspring of the same dam which hatched during that hatching cluster as one single clutch, or, alternatively, define the date range in clutches as the width of the hatching season.

This approach clearly will not work for a species that shows a birth–flow model. However, analysing inter–birth intervals using different date ranges can be helpful in defining the fine line between clutches. For this, it needs to be assumed that females of wild species (as opposed to some domestic bird species) do not continuously produce fertile eggs during the year. The clutch interval would be the interval that is larger than the evaluated date range.

References

Agostinelli C, Lund U (2013) R package circular: circular statistics (version 0.4-7). https://r-forge.r-project.org/projects/circular/

Barlow J, Boveng P (1991) Modeling age-specific mortality for marine mammal populations. Mar Mamm Sci 7(1):50–65

Benirschke K (1984) The frozen zoo concept. Zoo Biol 3(4):325–328

Blomqvist L (2004) International pedigree book for snow leopards (*Uncia uncia*) dataset. Technical report, Helsinki Zoo, Helsinki. Updated to 31 Dec 2003

Bradley AJ (2003) Stress, hormones and mortality in small carnivorous marsupials. In: Jones ME, Dickman CR, Archer M (eds) Predators with pouches: the biology of carnivorous marsupials. CSIRO Publishing, Collingwoord, Victoria, pp 254–267

Brouwer K, Schifter H, Jones ML (1994) Longevity and breeding records of ibises and spoonbills *Threskiornithidae* in captivity. Int Zoo Yearb 33:94–102

Caughley G (1966) Mortality patterns in mammals. Ecology 47(6):906–918

Caughley G (1977) Analyses of vertebrate populations. The Blackburn Press, Caldwell

Clarke AG (2009) The Frozen Ark Project: the role of zoos and aquariums in preserving the genetic material of threatened animals. Int Zoo Yearb 43:222–230

Cockburn A, Legge S, Double MC (2002) Sex ratios in birds and mammals: can the hypotheses be disentangled? In: Hardy ICW (ed) Sex ratios: concepts and research methods. Cambridge University Press, Cambridge, pp 266–286

Dijkstra C, Daan S, Buker JB (1990) Adaptive seasonal variation in the sex ratio of kestrel broods. Funct Ecol 4:143–147

Estes RD (2012) The behavior guide to African mammals: including hoofed mammals, carnivores, primates, 20th edn. University of California Press, Berkeley

Faust LJ, Bergstrom YM, Thompson SD, Bier L (2012) PopLink version 2.4. Lincoln Park Zoo, Chicago

Faust LJ, Thompson SD, Earnhardt JM, Brown E, Ryan S, Sherman M, Yurenka M (2003) Using stage-based system dynamics modeling for demographic management of captive populations. Zoo Biol 22:45–64

Flesness NR, Mace GM (1988) Population databases and zoological conservation. Int Zoo Yearb 27:42–49

Flesness NR, Lukens DR Jr, Porter SB, Wilson CR, Grahn LV (1995) ISIS and studbooks: very high correlation for the North American zoo population – a reply to Earnhardt, Thompson and Willis. Zoo Biol 14(6):509–517

Hurlbert SH (1984) Pseudoreplication and the design of ecological field experiments. Ecol Monogr 54(2):187–211

Jones ML (1993) Longevity of ungulates in captivity. Int Zoo Yearb 32:159–169

Kingdon J (2015) The Kingdon field guide to African mammals, 2nd edn. Bloomsbury Publishing, London

Komdeur J (2012) Sex allocation. In: Royle NJ, Smiseth PT, Kölliker M (eds) The evolution of parental care. Oxford University Press, Oxford, pp 171–188

Komsta L (2011) Outliers: tests for outliers. http://CRAN.R-project.org/package=outliers. R package version 0.14

Lacy RC, Ballou JD, Princée F, Starfield A, Thompson EA (1995) Pedigree analysis for population management. In: Ballou JD, Gilpin M, Foose TJ (eds) Population management for survival and recovery: analytical methods and strategies in small population conservation. Methods and cases in conservation science series. Columbia University Press, New York, pp 57–75

Lacy RC, Borbat M, Pollak JP (2009) VORTEX: a stochastic simulation of the extinction process, version 9.96. Technical report, Chicago Zoological Society, Brookfield

Lombardi CM, Hurlbert SH (1996) Sunfish cognition and pseudoreplication. Anim Behav 52: 419–422

McNutt JW, Silk JB (2008) Pup production, sex ratios, and survivorship in African wild dogs, *Lycaon pictus*. Behav Ecol Sociobiol 62:1061–1067

Morbey YE, Brassil CE, Hendry AP (2005) Rapid senescence in Pacific salmon. Am Nat 166(5):556–568

Mysterud A, Røed KH, Holland Ø, Yoccoz NC, Nieminen M (2009) Age-related gestation length adjustment in a large iteroparous mammal at northern latitude. J Anim Ecol 78(5): 1002–1006

Pewsey A, Neuhäuser M, Ruxton GD (2013) Circular statistics in R. Oxford University Press, Oxford

Princée FPG (2014) Population Management Library (PML) software and manual version 1.0. East Rudham. http://www.princee.com

R Core Team (2015) R: a language and environment for statistical computing. R Foundation for Statistical Computing, Vienna. https://www.R-project.org/

Scobie P, Bingaman Lackey L (1997–2012) Single Population Animal Record Keeping System (SPARKS) versions 1.42–1.66. International Species Information System, Eagan

Sokal RR, Rohlf FJ (2012) Biometry, 4th edn. W.H. Freeman and company, New York

Temte JL (1993) Latitudinal variation in the birth timing of captive California sea lions and other captive North Pacific pinnipeds. Fish Bull 91(4):710–717

Thompson SD, Earnhardt JM, Turner-Erfort G (1997) Guidelines for data entry and maintenance of North American regional studbooks. Lincoln Park Zoo, Chicago

Tukey JW (1977) Exploratory data analysis. Behavioural science: quantitative methods Addison-Wesley, Reading

Valenzuela N (2004) Evolution and maintenance of temperature-dependent sex determination. In: Valenzuela N, Lance V (eds) Temperature sex determination in vertebrates. Smithsonian Books, Washington DC, pp 131–147

Valenzuela N (2008) Sexual development and the evolution of sex determination. Sex Dev 2(2): 64–72

Warner D, Shine R (2008) The adaptive significance of temperature–dependent sex determination in a reptile. Nature 451:566–569

Weladji RB, Holand O (2003) Sex ratio in reindeer *Rangifer tarandus*: a test of the extrinsic modification hypothesis. Wildl Biol 9(1):29–36

Wiese RJ, Willis K (2004) Calculation of longevity and life expectancy in captive elephants. Zoo Biol 23:365–373

Zar JH (1984) Biological statistics, 2nd edn. Prentice-Hall, Englewood Cliffs

Chapter 4
Census Analysis

Abstract Direct census counts of wild populations depend on whether and when a species can be easily observed. Due to field conditions, the timing of census may not cover both sexes and/or all age groups. Census of sex and age groups relies on characteristics that can be observed under field conditions. These limitations do not apply to studbook census which is based on data from identifiable individuals. Timing of this virtual census is not bound by field conditions but can be optimised to the species' reproductive biology, e.g. pre– or post–breeding census, or classical end–of–year census. Census analysis of studbooks is generally based on annual intervals, but can be based on any interval that suits the life history of a species. The blesbok studbook is used as the main example. Graphic plots of census data visualise the history and dynamics of a captive population. This information is important when interpreting and validating results from the more complex life table analyses. Measures of population growth, i.e. finite rate of increase (λ) and intrinsic rate of increase (r), can be estimated from census data. Studbook software also estimates the geometric means of λ values for each year in retrospect. The annual values for λ are used to detect trends in population dynamic processes in the studbook population. The different methods that are used to express sex–ratio are described. Plotting sex–ratio at census dates is used to detect trends and shifts in the population.

4.1 Counting Animals

Census is defined as "complete counting of a whole population with respect to the variable under study" (Lawrence 2008). In the context of population biology this "variable" refers to the number of *living* individuals, at a given moment in time – the *census date*. In addition, individuals are categorised (whenever feasible) by sex and/or by life stage such as juvenile, sub-adult or adult. Census data can include a category for individuals of unknown sex. This category generally refers to new–borns whose sex has not been determined at the census date; and/or individuals of species whose sex may have to be determined at a later life–stage. Censuses that are repeated regularly over fixed time intervals (e.g. annually) provide information on trends in the population without knowledge of numbers of births, deaths, immigration and emigration. Moreover, census does not require identification of individuals.

F.P.G. Princée, *Exploring Studbooks for Wildlife Management and Conservation*,
Topics in Biodiversity and Conservation 17, DOI 10.1007/978-3-319-50032-4_4

Census is often considered in the context of counting animals in wildlife populations. The detail provided by census on wildlife populations depends on sexual dimorphism and life–stage related characteristics such as coat colour or shape of antlers/horns. These characteristics may not easily be observed in the field. Furthermore, the optimal period for census may differ between sexes. Therefore, it is not uncommon that census of wildlife populations is restricted to counting females as the reproductive unit.

Wildlife census is often envisioned as counting population numbers of species such as plains zebras (*Equus burchellii*) and buffalos (*Syncerus caffer caffer*) on African savannas. It is actually better to state "estimating numbers", as studies show that even counting savanna elephants (*Loxodonta africana*) is subject to underestimation (Jachman 1991). Estimating population sizes of species in rainforests often needs to be based on indirect signs such as dung piles. Transforming such indirect signs to actual number of individuals requires information on daily defaecation rates and decay rates (Buckland et al. 2001; Laing et al. 2003). However, these rates can vary per season and habitat and often require assumptions, given that species such as forest elephants (*Loxodonta cyclotis*) cannot readily be followed to measure defaecation rates (see Hedges 2012).

Difficulties in census are not necessarily restricted to wildlife populations. Species management in zoos also includes species that live in social groups, where individual identification is not always feasible (Mace et al. 1998; Princée 1995, 1998). This means that life table analysis cannot be applied. The same restrictions apply to groups that are managed under semi–free range conditions or have been reintroduced. Furthermore, the number of individuals in large zoo colonies of some species, such as bats (Chiroptera), may have to be extrapolated through indirect methods. For example, the colony size of the Egyptian fruit bat (*Rousettus aegyptiacus*) in Rotterdam Zoo is estimated from data on the actual daily food intake (G. Visser, personal communication).

Wildlife managers and ecologists are confronted with the difficulties of obtaining detailed data on wild populations. It is not surprising therefore that various statistical techniques have been developed to interpret limited data sets and to correct biased field data, and to provide decision tools for wildlife management (see e.g. Caughley 1977; Sinclair et al. 2006; Skalski et al. 2005). Management of studbook populations with either almost "perfect" data, or those populations that resemble characteristics of wildlife populations, can benefit from experiences with interpretation and handling of data from wildlife census. This benefit is reciprocal as census analysis on studbook populations with complete data sets can be used for testing "wildlife" census techniques. The European studbook for the blesbok, a species kept in harem groups, is used as the main example in this chapter (see background information in Chap. 2).

4.2 Census Date

Zoo inventory reports traditionally use 31 December as the annual census date. Census reports generated by the first studbook software also used this date (Princée 1989; Scobie and Flesness 1989). The "end–of–year" count may be appropriate for studbook analyses of species that have birth–flow breeding systems. However, it is recommended to adapt the census date for species which have clear birth–pulse breeding systems.

The choice of census date is either a date just before (*pre–breeding census*) or just after the breeding season (*post–breeding census*) (Caswell 2001). In the field, the choice depends on the best period to observe a species, females in particular. Visibility of individuals is not an issue, nor date or number of censuses, in studbook populations.

Census in studbook populations is carried out in retrospect i.e. after the studbook has been updated. Analysis software enables a census to be carried out at any date, or multiple dates e.g. both pre–breeding and post–breeding.

Pre–breeding census provides counts on living individuals with the youngest group born in the previous season (yearlings in annual census). This census type, therefore, provides information on the reproductive potential of the population. Post–breeding census, which includes new–borns, is not strictly necessary as studbook software counts all individual births per season (see Chap. 5).

Blesbok in the European region show a clear season of birth in June. The pre–breeding census date could be set to 14 June (see Table 3.6). The last census date used in the analysis is 14 June 2007.

4.3 Census

Graphic representation of census data provides a tool to visualise trends in studbook populations. Figure 4.1 shows census counts for males, females and the total population in the European studbook population of blesbok from the first historical (census) date in 1950 until 2007. This studbook population varied between 1 and 3 individuals from 1950 to 1960. It is very likely that these census data do not reflect the "true" historical population sizes as not every institution maintained a detailed record keeping system.

The period 1960–1970 is characterised by a slow increase. This still refers to the period before computerised in–house zoo registration became common. Therefore, this period may also not reflect the "true" historical population sizes. Census–style analyses such as number of imports, births and neonatal deaths can provide additional information by which to determine the level of completeness of data in a studbook (see Chap. 5). For example, stillbirths and individuals that died as neonates were not always registered in the past. In the case of the blesbok, the available data

Fig. 4.1 Annual census of the European zoo population of blesbok from start until 2007. The census date was set to 14 June (see text)

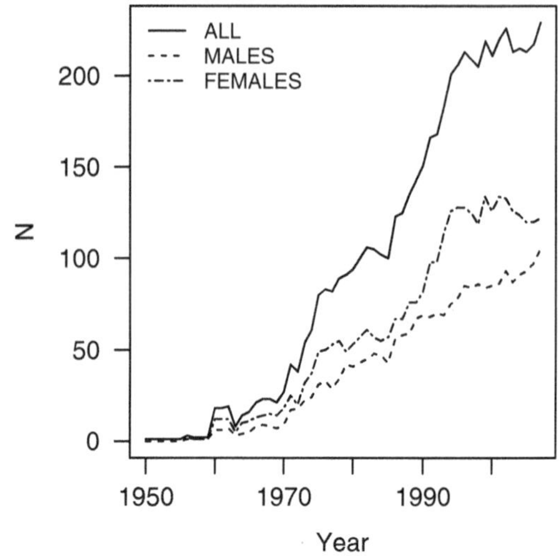

on neonatal mortality indicate that all individuals which died before contributing genetically to the population are included in the studbook.

Census analyses show the development of a population size over time. It is good practice initially to include all historical studbook data. It allows the determination of whether early data reflect the "true" population size or only comprise founders[1] and their descendants that have contributed genetically to the current studbook population (see Sect. 3.2.4).

The blesbok population shows a steady growth from 1970 until 1994 (Fig. 4.1). Since then the population is growing – with fluctuations – less rapidly; and the female population is actually declining. Figure 4.1 shows that females outnumber males at census dates. This reflects the management of this species in harem groups. However, the female population size was more or less stable from 1994, while the number of males was still increasing. The population was approaching equal sex–ratio in 2007.

Census analyses alone will not provide a causal explanation for the trend as observed in Fig. 4.1. For example, shifts in sex–ratio at census could be the result of shifted sex–ratio at birth and/or death (see Sect. 5.2.1). These trends could, for example, also reflect the result of management programmes. Fewer and smaller harem groups may be maintained to control population growth (which would be in line with the reduced population growth), and bachelor herds may have been established to maintain potential genetic variation in the population.

Since this census analysis involves a regional (European) zoo population, import and export to other regions (see Sect. 5.4) may also affect the differences in

[1] Founders are wild–born individuals that have produced (viable) offspring.

sex–ratio. These questions partly arise due to analysing this studbook without prior knowledge of management measures that could be in place. This stresses the importance of working together with studbook keepers and/or breeding programme managers in the interpretation of analyses.

4.4 Population Growth

It is important to describe and interpret the historical and current patterns (trends) as observed in raw census data (see Fig. 4.1). This provides a first feel for processes that are occurring or have occurred in the population. The next step is to quantify these patterns in terms of the *finite rate of increase* (symbol: λ) between census intervals (Caughley 1977):

$$\lambda = N_{t+1}/N_t \tag{4.1}$$

where N_t and N_{t+1} are the population sizes at census dates t and $t+1$, respectively.

A value of $\lambda > 1$ means an increase in population size; $\lambda = 1$ means no growth; $0 < \lambda < 1$ means a decrease in population size. Values for growth rates are often presented as percentages: $(\lambda - 1) \times 100$ (Caughley 1977). Thus values for λ of 110 % (1.1) and 90 % (0.9) refer to a 10 % increase and decrease between intervals, respectively.

Figure 4.2 shows annual growth rates in the European studbook population of blesbok between census dates 1970–2007. Growth rates can, as shown in this figure, fluctuate greatly between censuses, especially in the initial phase after a

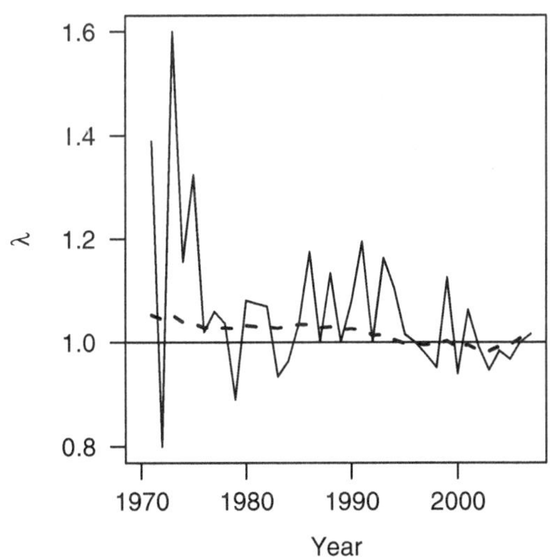

Fig. 4.2 Annual growth rates in the European studbook population of blesbok between census dates 1970–2007. The census date was set to 14 June. The dashed line refers to the geometric mean λ_G. The solid line represents a stable population size, i.e. $\lambda = 1$

studbook population has been established. Therefore, averaging λ would provide more meaningful values. The geometric mean is often used in averaging ratios and percentage changes (Zar 1984), and has been applied to population growth rates (Shryock et al. 1980; Woofter 1932).

The programs SPARKS (Scobie and Bingaman Lackey 1997–2012), PopLink (Faust et al. 2012) and PML (Princée 2014) implement geometric means of λ in their census analyses. These means are calculated for every period before the last census (see Eq. 4.2). For example, the geometric mean listed for the year 1971 refers to all λ values starting in 1971 to the last census. This means that the geometric mean changes in retrospect, i.e. each time a new census has been carried out. This is shown in Eq. 4.2.

$$\lambda_G(t) = antilog \frac{1}{n} \sum_{t}^{t-n} log\, \lambda \tag{4.2}$$

where n is the number of censuses previous to the last census date t and $n > 2$.

The dashed line in Fig. 4.2 represents the geometric mean of λ for each year. Table 4.1 shows geometric means for the year 1971 and the period 2004–2007.

The geometric mean clearly "smooths" the peaks in annual growth rates (Fig. 4.2). For example, the growth rate between 1970 and 1971 is 1.556 (55.6 %) while the geometric mean is 1.059 (or 5.9 %). Growth rates for the period 2003–2007 are in the range 0.942–1.055, which results in a geometric mean of 1.003 (see Table 4.1).

The finite rate of increase (λ) is a relatively easy measure for interpreting trends in populations within the context of "human" time–units such as a year. However, the *intrinsic or instantaneous rate of increase* (symbol: r) is the preferred measure in ecology (Caughley 1977).

Table 4.1 Annual growth rates (λ) and geometric means (λ_G) in the European studbook population of blesbok for the years 1970–1972 and 2003–2007

Year	N	λ	λ_G	Period
2007	229	1.055		
2006	217	1.019	1.037	2006–2007
2005	213	0.991	1.021	2005–2007
2004	215	1.009	1.018	2004–2007
2003	213	0.942	1.003	2003–2007
....	...	.....	.....	
1972	38	0.905	1.048	1972–2007
1971	42	1.556	1.059	1971–2007
1970	27			

N population size, λ annual growth rate, λ_G geometric mean of annual growth rates

The relation between λ and r is:

$$\lambda = e^r \tag{4.3a}$$

$$r = \ln \lambda \tag{4.3b}$$

where $e = 2.718282$ (base of natural logarithms) (Caughley 1977).

To illustrate the conversion between both measures: values for λ of 0.5, 1 and 2 result in the r values -0.69, 0 and 0.69, respectively. Thus a negative value of r refers to population decrease, whereas no growth (i.e. stable size) is indicated by $r = 0$. The measure r can be more easily converted to other time–units than λ. For example, if r per year is x, than r per day is $x/365$.

The measure r is often pronounced as "Little r" to avoid confusion with "Big R", the *net reproductive rate* (symbol: R_0), which is the population growth per generation time (symbol: $\bar{T}$) (see Chap. 9).

4.5 Sex–Ratio

Sex–ratio is typographically presented as *males* : *females* e.g. 1 : 1, 1 : 4, 50 : 50 or the actual number of males to the number of females i.e. $N_m : N_f$.

As females are the reproductive unit, the measure of sex–ratio N_f/N_m is generally used in wildlife management (Caughley 1977; Skalski et al. 2005). However, this measure has the disadvantage that the function is asymptotic to the right as division by zero (males) is undefined. This problem does not exist when sex–ratio is expressed as the proportion of females:

$$sex\ ratio = \frac{N_f}{N_f + N_m} \tag{4.4}$$

where N_f and N_m are the numbers of females and males, respectively.

Another advantage of using proportions is that these values can be interpreted as probabilities in stochastic population models such as VORTEX (Lacy et al. 2009).

Figure 4.3 presents sex–ratio (as proportion of females) in the European zoo population of blesbok at census dates for the period 1970–2007. Although this figure does not provide new information additional to plotting census data of both males and females (see Fig. 4.1), it better shows shifts and trends in sex–ratio in the population.

Sex–ratio at census date is the result of sex–ratios in births, deaths and migration. These population dynamics in captive populations are presented and discussed in Chap. 5. The topic of sex allocation is discussed in the context of sex determination and sex–ratio at birth.

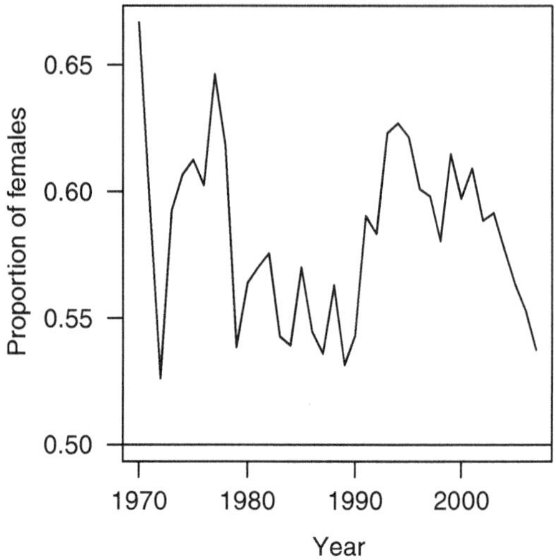

Fig. 4.3 Sex–ratio at census date in the captive population of blesbok in European zoos from 1970 to 2007. Pre–breeding census date is 14 June

References

Buckland ST, Anderson DR, Burnham KP, Laake JL, Borchers DL, Thomas L (2001) Introduction to distance sampling: estimating abundance of biological populations. Oxford University Press, New York

Caswell H (2001) Matrix populations models: construction, analysis, and interpretation, 2nd edn. Sinnauer Associates, Sunderland

Caughley G (1977) Analyses of vertebrate populations. The Blackburn Press, Caldwell

Faust LJ, Bergstrom YM, Thompson SD, Bier L (2012) PopLink, version 2.4. Lincoln Park Zoo, Chicago

Hedges S (ed) (2012) Monitoring elephant populations and assessing threats. Universities Press, Himayatnagar

Jachman H (1991) Evaluation of four survey methods for estimating elephant densities. Afr J Ecol 29(3):188–195

Lacy RC, Borbat M, Pollak JP (2009) VORTEX: a stochastic simulation of the extinction process, version 9.96. Technical report, Chicago Zoological Society, Brookfield

Laing SE, Buckland ST, Burns RW, Lambie D, Amphlett A (2003) Dung and nest surveys: estimating decay rates. J Appl Ecol 40:1102–1111

Lawrence E (ed) (2008) Henderson's dictionary of biology, 14th edn. Pearson Education, Harlow

Mace G, Hall H, Vedmar A (eds) (1998) Population group management workshop, Zoological Society of London, London, 5–7 Oct 1998

Princée FPG (1989) Zooresearch studbook management program, version 1.03, software and manual. National Foundation for Research in Zoological Gardens, Amsterdam

Princée FPG (1995) Overcoming the constraints of social structure and incomplete pedigree data through low-intensity genetic management. In: Ballou JD, Gilpin M, Foose TJ (eds) Population management for survival and recovery: analytical methods and strategies in small population conservation. Methods and cases in conservation science series. Columbia University Press, New York, pp 124–154

Princée FPG (1998) Genetic management of small animal populations in zoos and wildlife reserves: the use of genome models in the estimation of genetic variation and the effects of social structures. PhD thesis, University of Groningen, Groningen

Princée FPG (2014) Population Management Library (PML) software and manual, version 1.0. East Rudham. http://www.princee.com

Scobie P, Bingaman Lackey L (1997–2012) Single Population Animal Record Keeping System (SPARKS), versions 1.42–1.66. International Species Information System, Eagan

Scobie P, Flesness NR (1989) Single Population Animal Record Keeping System (SPARKS), software and manual. International Species Information System, Apple Valley

Shryock HS, Siegel J (1980) U.S. Bureau of the census – the methods and materials of demography, fourth printing (rev.) edn. U.S. Government Printing Office, Washington, DC

Sinclair ARE, Fryxell JM, Caughley G (2006) Wildlife ecology, conservation and management, 2nd edn. Blackwell Publishing, Malden

Skalski JR, Ryding KE, Millspaugh JJ (2005) Wildlife demography: analysis of sex, age and count data. Elsevier Academic Press, Burlington

Woofter TJ (1932) Interpolation of populations whose rate of increase is declining. J Am Stat Assoc 27(178):180–182

Zar JH (1984) Biological statistics, 2nd edn. Prentice-Hall, Englewood Cliffs

Chapter 5
Births, Deaths and Migration

Abstract Data on numbers of births, deaths, emigrants and immigrants between census dates are required properly to understand population dynamics as observed in census counts. Such data can be difficult to obtain for wild populations, but are relatively easy to obtain from studbooks. Crude birth rates that are estimated from birth and census counts provide an idea about the reproductive potential of the population. This measure can be refined, for example to female births per adult female. This chapter discusses several topics that are related to biased sex–ratio at birth. The different sex determination systems in the animal kingdom and their potential effects on biased sex–ratio are discussed. Equality in sex–ratio per breeding season or over longer periods of time can be tested with the binomial test. The annual data on sex–ratio in births provide material to study sex allocation at the (studbook) population or individual institutional level. The runs test can be used to detect sex allocation in annual data. Patterns of annual death rates and the proportion of neonatal (30 day) mortality can be obtained from death and census counts. The last section presents migration, which refers to imports from the wild and reintroduction within the context of a single studbook population, or transfers between any population in the wider context of metapopulation management, e.g. transfers between regional zoo populations, between zoos and with reintroduced populations.

5.1 Census Events

Census analysis provides information on population development, but does not provide detail on the underlying processes. Data on numbers of births (symbol: B), deaths (symbol: D), emigrants (symbol: E) and immigrants (symbol: I) *between* census dates are required to understand population dynamics properly. These processes can be "captured" in a simple but fundamental population model:

$$N_{t+1} = N_t + (B_{[t,t+1]} - D_{[t,t+1]}) + (I_{[t,t+1]} - E_{[t,t+1]}) \tag{5.1}$$

where t and $t + 1$ are census dates (Skalski et al. 2005).

Collecting data on these dynamic processes (*census events*) in wildlife populations (including semi–free and reintroduced populations) will be subject to similar

F.P.G. Princée, *Exploring Studbooks for Wildlife Management and Conservation*, Topics in Biodiversity and Conservation 17, DOI 10.1007/978-3-319-50032-4_5

practical difficulties as described for census data (Chap. 4). Counting the number of births may be feasible in species with a distinct birth season (birth–pulse model) where census takes place at a post–breeding date. Deaths and migration will take place throughout the year. These census events not only require regular surveys, but are also more difficult to assess.

In general, death due to predation is rarely witnessed in wild populations. Whereas skulls of African hoofstock on savanna can be recovered (e.g. Spinage 1972), the situation in rainforest is more difficult. For example, even the carcass of a forest elephant that is some metres from a track is not always detected (A. Turkalo, personal communication). Determining which individuals are immigrants depends entirely on being able to identify individuals.

Studbooks will provide more than sufficient detail to determine population parameters such as births, deaths and migrants. Even estimated dates can be included in analyses, as long as the event took place within the census interval, e.g. the year of birth or death is sufficient for annual censuses at the end of the year. The European studbook for the blesbok is used as the main example in this chapter.

5.2 Births

Censuses of number of births between census intervals provide historical information regarding breeding success and/or failure in the studbook population. Figure 5.1 shows the number of births in the European studbook population of blesbok from 1970 to 2007. Breeding in this population started from 1970 on and increased steadily after that. Combining census counts of births with those of imports (Sect. 5.4) provides information about when the captive population became

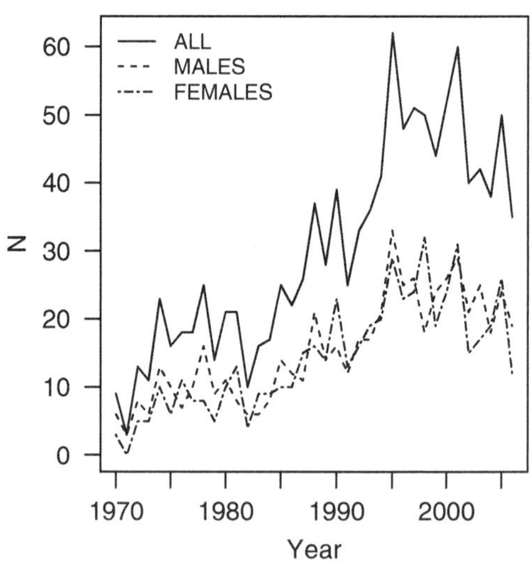

Fig. 5.1 Annual number of births in the captive population of blesbok in European zoos from 1970 to 2007. Pre–breeding census date is 14 June

independent of imports. This allows fine–tuning in the selection of relevant time periods in demographic analyses to evaluate captive populations (Faust et al. 2003).

Since birth census provides only absolute numbers, this is insufficient to gauge whether reproduction is enough to maintain the population. Therefore, birth data need to be combined with census data on living individuals (Chap. 4) in order to calculate birth rates.

The *crude birth rate* (symbol: b_t), between census dates is computed as:

$$b_t = B_{[t,t+1]}/N_t \tag{5.2}$$

where $B_{t,t+1}$ is the number of births between census dates t and $t + 1$, and N_t is population size at census date t. The crude birth rate is also expressed as the number of births per 100 or per 1,000 individuals (of the sex and age groups in the census).

Figure 5.2a presents crude birth rates between annual censuses for the captive blesbok population from 1970 to 2007. A similar pattern of large fluctuations as discussed for annual growth rates (see Sect. 4.4) can be observed in births. The median inter–birth interval of 388 days suggests that this species is an annual breeder. The geometric means as calculated according to Eq. 4.2 "smooth" the annual crude birth rates results. These mean values fluctuate around a rate of 0.2 (or 20 births per 100 individuals) until 1995, but then gradually decline. The decline in crude birth rate is expected from the slower population growth in the recent period (see Fig. 4.1). However, the decline in population growth is not determined by births alone, but also by mortality and migration (see Sects. 5.3 and 5.4).

Wildlife managers often only consider the female population, as being the reproductive unit, in censuses. Furthermore, it is often easier to count females (with

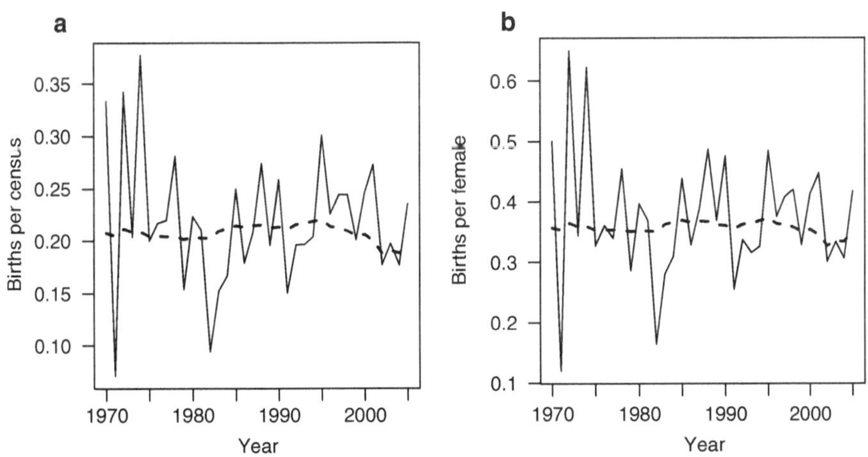

Fig. 5.2 (**a**) Crude birth rates in the captive population of blesbok in European zoos from 1970 to 2007. Pre–breeding census date is 14 June. The *dashed line* refers to the geometric mean. (**b**) Births per female (natality or productivity)

offspring) than males (Caughley 1977). The crude birth rate then refers to the ratio births per female (symbol: b_{Ft}) under such a census regime. The ratio births per female is also named *natality* (Krebs 1994) or *productivity* (Skalski et al. 2005). Figure 5.2b presents the annual productivity in the blesbok population. Geometric means are, as expected, higher and fluctuate around a birth rate of 0.35.

Following the idea of females being the reproductive unit, one can add a refinement by calculating the ratio female births per female (symbol: b_{FFt}), named *fecundity* (Skalski et al. 2005). These rates will be on average half the values observed in total births per female when sex–ratio at birth is equal (see Sect. 5.2.1). Fecundity rates are presented for the blesbok population in Fig. 5.3a.

Fecundity and productivity actually refer to census of females in the reproductive age group (Krebs 1994; Skalski et al. 2005). The reproductive life–span of female blesbok in European zoos ranges from 719–5,193 days ($\approx$2–14 years) in the period 1970–2007 ($n = 601$, 1 % truncation). Fecundity rates that take these reproductive ages into account are presented in Fig. 5.3b. These rates are considerably higher than those in Fig. 5.3a as they refer to females that can potentially breed. This method may not be feasible for wildlife populations of species where sex–determination of newborn and/or juveniles and/or age determination at census date is difficult.

The choice of census date is important for species with a birth–pulse system. Pre–breeding census dates are recommended as these counts will reflect the number of potential breeders close to the birth season.

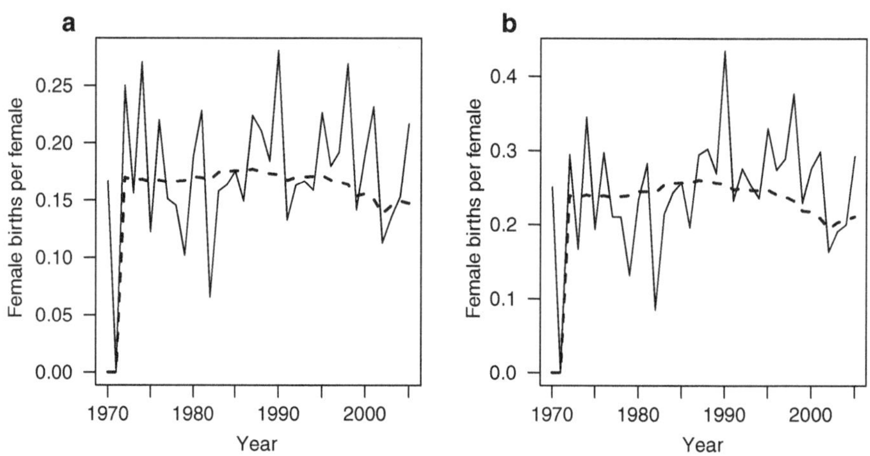

Fig. 5.3 (**a**) Female births per female in all age groups (fecundity) in the captive blesbok population in European zoos from 1970 to 2007. The census date was set to 14 June. The *dashed line* refers to the geometric mean. (**b**) Female births per sexually mature female (fecundity)

5.2.1 *Sex–Ratio at Birth*

Figure 5.4 presents sex–ratio at birth in the blesbok studbook population between 1970 and 2007. The sex–ratio fluctuates annually, with an extreme of no female births in 1971. Since only three births occurred in 1971, the chance of having all males, assuming that chances to be born as male or female are equal, is not that extreme i.e. $p = 0.125$. Nevertheless, the 1971 data could be considered as outliers due to the small number of births, whenever annual trends in sex–ratio are studied. However, these data could be included when evaluating sex–ratio in the total number of births that occurred in the period 1970–2007.

The mean sex–ratio at birth of 0.483 ($B_m = 570$, $B_f = 533$) suggests a small male bias. However, the null hypothesis, i.e. sex–ratio at birth is equal, is not rejected in a binomial test (R Core Team 2015; Zar 1984) ($p = 0.278$).[1]

Sex–ratio at birth is not necessarily equal in all populations, and can have serious practical impact on population management (Faust and Thompson 2000; Glatston 1997). The mechanisms and processes behind biased sex–ratios will be briefly discussed in the following sections.

5.2.1.1 Sex Determination

Sex determination systems in the animal kingdom vary between strict genetic control to strict environmental control. Sex chromosome systems such as the male

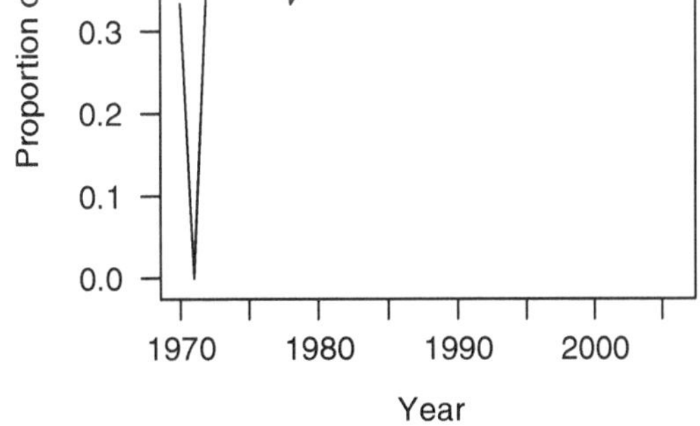

Fig. 5.4 Sex–ratio at birth in the captive population of blesbok in European zoos (1970–2007)

[1]Sex at birth could not be determined in 19 individuals. The null hypothesis is also not rejected in the case that these individuals are assumed to be either all male or female.

XX/XY system in mammals[2] and the female ZZ/ZW system in birds are examples of genotypic (heterogametic) sex determination (GSD) systems that result in sex determination at fertilisation (Valenzuela 2008). The chances of being male or female in a GSD system are expected to be equal, resulting in a sex–ratio at fertilisation that is expected to follow a binomial distribution.

Environmental sex determination (ESD) takes places during embryogenesis and has been observed among vertebrates in fish, amphibians and reptiles (Devlin and Nagahama 2002; Valenzuela 2008). Temperature–dependent sex determination (TSD) during incubation has been extensively studied in reptiles (see overviews Ciofi and Swingland 1997; Valenzuela 2004). The TSD system has been observed in tuataras, all studied crocodile species, it is prevalent in turtles, common in lizard species of the families Agamidae and Geckonidae, but has not been observed in snakes.

The temperature effects differ among reptile taxa, e.g. low temperatures result in females in most lizards, males in many chelonians. A third TSD type that is found in many crocodilians results in females at low and high temperatures, and males at intermediate temperatures (see Ciofi and Swingland 1997). Other environmental factors, such as the pH, are known to influence sex–ratio in fish species (see overview in Devlin and Nagahama 2002).

Reptile clutches that are laid under captive conditions are generally moved to incubators where temperature and humidity can be controlled. This provides opportunities to manipulate sex–ratio in species that show the TSD system. For example, if too many males have been born in recent years, the studbook manager may recommend incubation temperatures that result in females to achieve equal sex–ratio. However, the sex determination system is not fully known, especially in fish species. Analysis of sex–ratio at birth in studbook populations is useful, especially in combination with information on incubation conditions, in providing more detail on the underlying sex determination system (Ciofi and Swingland 1997).

5.2.1.2 Biased Sex–Ratio

The stochastic nature of GSD systems explains fluctuations in annual sex–ratio in small populations, such as the blesbok studbook population where annual births in the study period do not exceed 60 (see Figs. 5.1 and 5.4). Although overall sex–ratio over a longer period of time is expected to be equal, male or female bias has been observed in birds and mammals (Hardy 2002; Komdeur 2012).

Some of the studbook examples in this book also show biased sex–ratios. For example, the sex–ratio at birth in the International studbook population of mountain bongo (total number of births (B) for the years 1977–2008) is female biased i.e. 0.55 ($B = 2{,}045, p < 0.001$). A significant male biased sex–ratio of 0.46

[2]Some exceptions e.g. the platypus exist.

($B = 1{,}115$, $p = 0.02$) is observed in the European studbook population of African wild dogs for the years 1986–2012 (which happens to be similar to the observed sex–ratio in a wild population in Northern Botswana (McNutt and Silk 2008)).

5.2.1.3 Sex Allocation

Given the large numbers of births in the studbook populations of African wild dogs and lowland bongo, the observed biases cannot be explained by stochastic effects. The explanation maybe found in *sex allocation*, which refers to the resources that are allocated to male versus female offspring (Fisher 1930).

Sex allocation in birds and mammals is the subject of scientific debates (see overviews in e.g. Cockburn et al. 2002; Frank 1990; Komdeur 2012). It is beyond the scope of this book to discuss this subject in great detail. However, the practical impact of biased sex–ratio, and the potential that studbook data can contribute to science (e.g. Faust and Thompson 2000; Glatston 1997; Thogerson et al. 2013) merits some attention to this subject.

Five main categories of adaptive models that explain sex–ratio in birds and mammals are recognised (Cockburn et al. 2002). One of the basic models is *Fisher's equal allocation theory* (Fisher 1930), which states that selection should favour an unbiased sex ratio at the population level. The *homeostasis theory* is a variant in which the parents produce the sex that is rare (Cockburn et al. 2002).

Equal mean sex–ratio does not exclude effects of sex allocation. Nonrandom patterns such as years of male bias that are followed by years of female bias can indicate sex allocation. Figure 5.5 shows years with statistically significant ($p < 0.05$) female bias ("F") and male bias ("M") in sex–ratio in Nepalese red pandas for the period 1990–2010. The mean sex–ratio for this period is exactly 0.5 ($B_m = B_f = 652$).

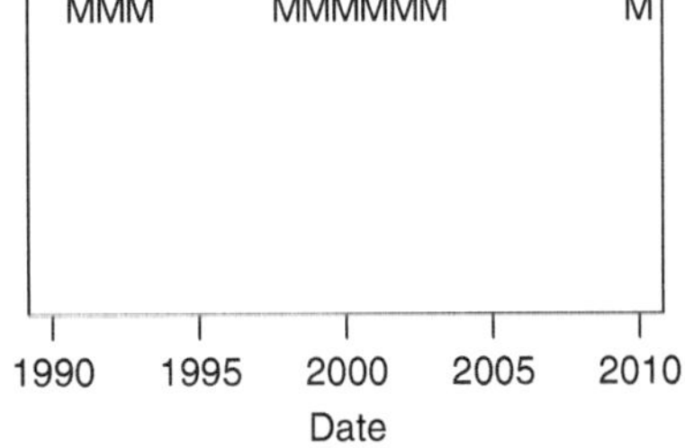

Fig. 5.5 Bias in sex–ratio at birth in the captive population of Nepalese red pandas (1990–2010). *F* = female bias, *M* = male bias, the *dotted line* represents sex–ratio. Sex–ratio per year is tested for significant difference from equality using a binomial test ($\alpha = 0.05$)

Several sequences of male and female bias can be observed during this period. The *runs test* can be applied to test the randomness of these sequences ("runs") (Sokal and Rohlf 2012). The graph in Fig. 5.5 can also be represented as text: "MMMFFFMMMMMMFFFFM". Years of equal sex–ratio are omitted in the runs test. The null hypothesis that the above sequences are random is rejected ($p = 0.028$) by a two–sided runs test in the R package *tseries* (Trapletti and Hornik 2015).

Runs tests require dichotomous data, and cannot, therefore, be applied to data that are sex biased for one sex. However, for such biased data–sets with more than 5 births, the binomial test will reject (i.e. $p < 0.05$) the null hypothesis that the sex–ratio is equal anyway.

The above example illustrates bias in sex–ratio at the population level. However, sex allocation can occur at the individual level as well. The *Trivers–Willard effect* (Trivers and Willard 1973) refers to polygynous species in which females in good condition are expected to produce (more) sons. Studies for this model require pedigree data, such as available in studbooks. The chronological sequence(s) of sex of an individual's offspring are tested for randomness with the runs test (newborns of unknown sex are omitted).

The introduction of computerised in-house record keeping systems in the mid 1980s means that to date some 30 years of detailed historical data are available for most species in zoological gardens. Studbooks can even include data that go back to the early 1900s e.g. snow leopard (see Chap. 3). These data allow the study of (reproductive) fitness over multiple generations, e.g. a recent study using long–term data from San Diego Zoo tested whether parents ($P0$) with biased sex–ratio among their offspring, "produced" more grandchildren ($F2$) (Thogerson et al. 2013).

Zoo populations have been used in studies of sex allocation (Glatston 1997). These studies are also informative about difficulties that can be encountered in interpreting zoo data in terms of sex allocation theory. For example, the male bias in African wild dogs may refer to the sex–ratio around the time pups emerge from the den (see Chap. 3). Sex determination in various species cannot take place until newborns can be separated from the parents. The sex of those that die earlier, especially neonates, can often not be accessed. This means that the possibility that early mortality is sex–biased cannot be excluded.

Since environmental conditions in captivity differ from those in the natural habitat, mechanisms of sex allocation may differ or sex allocation may even be absent. It is clear, however, that biased sex–ratios do occur in zoo populations.

5.3 Deaths

Estimating mortality in wild populations depends on either finding remains (carcasses) or deducing numbers of deaths from census data, births and net migration (e.g. Caughley 1977; Spinage 1972). The situation for studbook populations is different as the date of death of most individuals is known. Even estimated dates, as long as the assumed deviation in time falls within a census interval, can be included in studbook analyses.

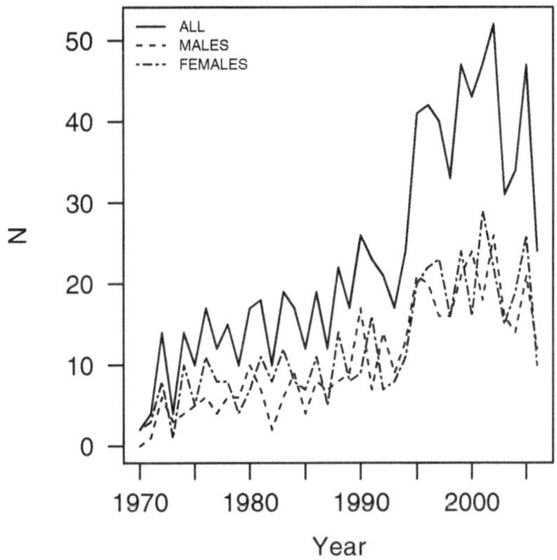

Fig. 5.6 Annual number of deaths in the captive population of blesbok in European zoos from 1970 to 2007. The census date is 14 June

Figure 5.6 presents the annual number of deaths in the blesbok population from 1970 to 2007. Annual deaths in females and males do not seem to differ dramatically at first glance. This figure provides an impression about the order of magnitude in numbers of annual deaths. However, trends need to be evaluated within the context of census data and/or number of births.

The *crude death rate* (symbol: d_t) between census dates is computed as:

$$d_t = D_{[t,t+1]}/N_t \tag{5.3}$$

where $D_{[t,t+1]}$ is the number of deaths between census dates t and $t+1$, and N_t is population size at census date t. Crude death rate is also presented as number of deaths per 100 or per 1,000 individuals.

Crude death rates in the blesbok population are presented in Fig. 5.7a. Death rates fluctuate annually, especially in the first years when the population was small. Geometric means of annual death rates (see Eq. 4.2) provide a "smoother" pattern that makes it easier to detect possible trends.

The geometric means range from 0.15 to 0.18 per year during the period 1970–2008 (Fig. 5.7a). Death rates in the period 1994–2001 are higher than in previous and later periods. However, one cannot draw conclusions regarding the reasons/causes of this trend without additional information.

The crude death rates in Fig. 5.7a are based on deaths in all age groups. Mortality in neonates and juveniles is high compared to other age groups (see Chap. 3). This means that an increase in births can result in an increase in (total) deaths (and rates) too. Figure 5.7b presents the annual proportion of neonatal deaths (deaths within 30 days) of total number of deaths. The pattern in neonatal mortality, however, does not fully explain the pattern in total death rates during the period 1994–2001.

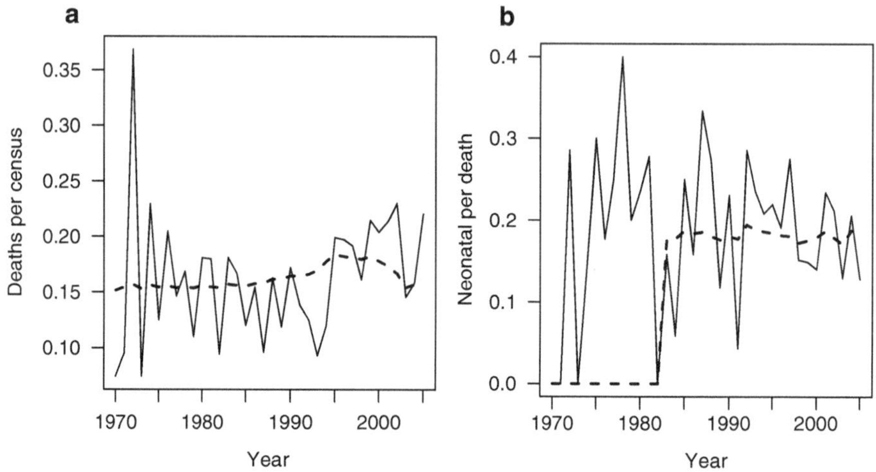

Fig. 5.7 (**a**) Crude death rates in the captive population of blesbok in European zoos from 1970 to 2008. Pre–breeding census date is 14 June. The *dashed line* refers to the geometric mean. (**b**) Proportion of neonatal deaths (deaths within 30 days) of total annual deaths

The absence of neonatal mortality until the mid 1980s is likely to be an artefact i.e. only (living) ancestors born in the pre–studbook period are registered. This artefact can explain the lower death rates observed in the period prior to 1994.

5.4 Migration

The term *migration* has different meanings within biological disciplines (Allendorf et al. 2013). Ecologists will generally refer to the movement of individuals during their lifetime from one geographic area or region to another e.g. seasonal migration of birds. Population geneticists refer to migration as movement of individuals from one (genetic) population to another.

Studbooks were established to genetically manage closed (sub–)populations, and so migration refers to planned transfers of individuals. Therefore using the "geneticist" definition of migration makes more sense in the context of studbook management. The relevance of migration in the context of population dynamics is to evaluate the effects of migration on population growth.

The term *dispersal*, which is generally used in ecological literature to refer to genetic exchange between populations (Allendorf et al. 2013; Caughley 1977), can also be applied within the context of studbook populations. Dispersal not only refers to exchanges between populations, but also to exchanges between social (breeding) groups within the same population as well (Princée 1995, 1998; Shields 1987).

5.4.1 Migration Patterns

Studbook management in the context of meta–populations stretches beyond captive populations. Therefore, the terms *immigration* and *emigration* are used instead of the traditional zoo terms import and export. Figure 5.8 presents a schematic representation of possible migration patterns in managed (zoo) studbook populations.

Immigration refers to import of individuals from the original wild population(s), transfers from other regional zoo populations and transfers from reintroduced populations; emigration refers to reintroduction and transfers to other regional zoo populations.

Figure 5.9a presents imports into the European zoo population of blesbok since 1970. The largest numbers of imports occurred in the period 1970–1975. These imports not only refer to wild–born individuals, but to individuals of unknown provenance as well. The total number of imports is important in determining whether a population is growing through births or imports. Imports of wild–born individuals, however, are of more interest for population genetics. Figure 5.9b shows that imports of blesbok from the wild population are considerably fewer, and infrequent, compared to the total number of imports.

Migration rates can provide a better insight into population dynamic processes, as discussed for births and deaths. The *immigration rate* (symbol: i_t) and *emigration rate* (symbol: e_t) are calculated as:

$$i_t = I_{[t,t+1]}/N_{[t]} \tag{5.4a}$$

$$e_t = E_{[t,t+1]}/N_{[t]} \tag{5.4b}$$

where $I_{[t,t+1]}$ is the number of immigrants and $E_{[t,t+1]}$ the number of emigrants between census dates t and $t+1$, respectively; and N_t is the census at date t.

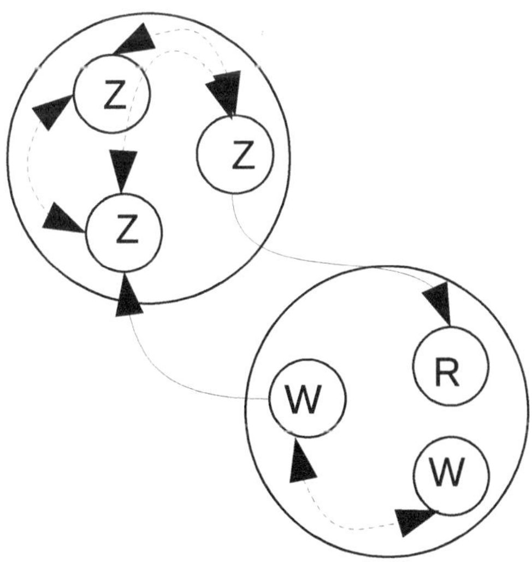

Fig. 5.8 Migration patterns in managed zoo populations. W = wild population, Z = zoo region, R = reintroduced population

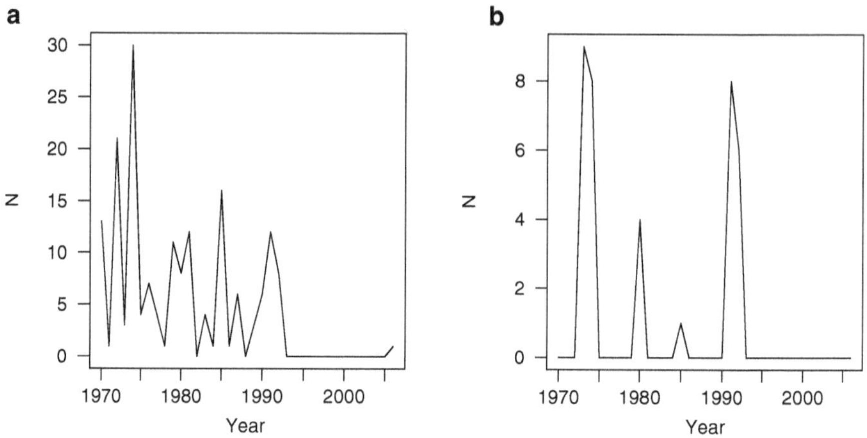

Fig. 5.9 (**a**) All imports in the captive population of blesbok in European zoos from 1970 to 2008. Pre–breeding census date is 14 June. (**b**) Imports from the wild

Fig. 5.10 Historic overview of census (*solid line*), births (○), deaths (×) and imports (◇) in the European studbook population of blesbok. The Y–axis is on a logarithmic scale

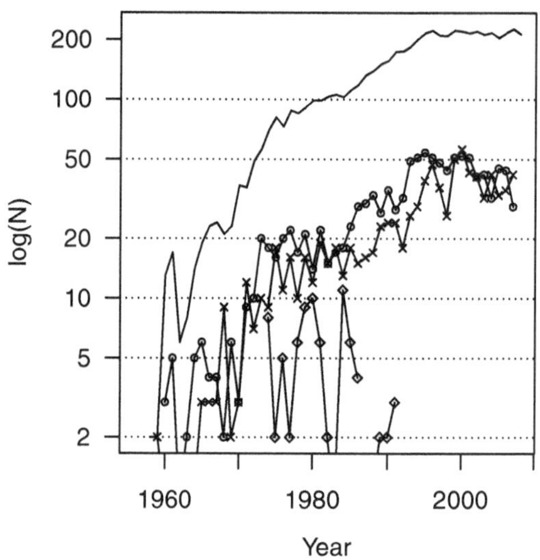

5.5 Bringing it Together

Equation 5.1 in the introduction of this chapter describes the relation between population dynamic processes (births, deaths and migration) and population size. Each of these processes has been described separately in the previous sections to show that census events provide more information than simply "being" numbers that make up population size. However, returning to population dynamics, it is good practice to bring all the census–style data of a studbook together in a single graph.

Figure 5.10 presents census (solid line), births (open circle), deaths (cross) and imports (diamond) in the blesbok population since 1959. The Y–axis is on a logarithmic scale in order to combine the relatively small numbers of imports with census data in a single graph.

This overview shows the effects of the different dynamic processes, in relation to each other, on population development, e.g. births that outnumber imports, births and deaths that tend to equalise, although still fluctuating, since 2000. As mentioned previously, historical overviews are important to decide which periods of time are of special interest to analyse, or whether sufficient data are available for more detailed analysis, e.g. for the life tables that will be discussed in Chap. 7.

References

Allendorf FW, Luikart G, Aitken SN (2013) Conservation and the genetics of populations, 2nd edn. Wiley–Blackwell, Chichester

Caughley G (1977) Analyses of vertebrate populations. The Blackburn Press, Caldwell

Ciofi C, Swingland IR (1997) Environmental sex determination in reptiles. Appl Anim Behav Sci 51(3–4):251–265

Cockburn A, Legge S, Double MC (2002) Sex ratios in birds and mammals: can the hypothesis be disentangled? In: Hardy (2002), pp 266–286

Devlin RH, Nagahama Y (2002) Sex determination and sex differentiation in fish: an overview of genetic, physiological, and environmental influences. Aquaculture 208(3–4):191–366

Faust LJ, Thompson SD (2000) Birth sex ratio in captive mammals: patterns, biases, and the implications for management and conservation. Zoo Biol 19:11–25

Faust LJ, Thompson SD, Earnhardt JM, Brown E, Ryan S, Sherman M, Yurenka M (2003) Using stage–based system dynamics modeling for demographic management of captive populations. Zoo Biol 22:45–64

Fisher RA (1930) The genetical theory of natural selection, 3rd edn. Clarendon Press, Oxford

Frank SA (1990) Sex allocation theory for birds and mammals. Ann Rev Ecol Syst 21:13–55

Glatston AR (1997) Sex ratio research in zoos and its implications for captive management. Appl Anim Behav Sci 51(3–4):209–216

Hardy ICW (ed) (2002) Sex ratios: concepts and research methods. Cambridge University Press, Cambridge

Komdeur J (2012) Sex allocation. In: Royle NJ, Smiseth PT, Kölliker M (eds) The evolution of parental care. Oxford University Press, Oxford, pp 171–188

Krebs CJ (1994) Ecology: the experimental analysis of distribution and abundance, 4th edn. Addison–Wesley, San Francisco

McNutt JW, Silk JB (2008) Pup production, sex ratios, and survivorship in African wild dogs, *Lycaon pictus*. Behav Ecol Sociobiol 62:1061–1067

Princée FPG (1995) Overcoming the constraints of social structure and incomplete pedigree data through low-intensity genetic management. In: Ballou JD, Gilpin M, Foose TJ (eds) Population management for survival and recovery: analytical methods and strategies in small population conservation. Methods and cases in conservation science series. Columbia University Press, New York, pp 124–154

Princée FPG (1998) Genetic management of small animal populations in zoos and wildlife reserves: the use of genome models in the estimation of genetic variation and the effects of social structures. PhD thesis, University of Groningen, Groningen

R Core Team (2015) R: a language and environment for statistical computing. R Foundation for Statistical Computing, Vienna. https://www.R-project.org/

Shields WM (1987) Dispersal and mating systems: investigating their causal connections. In: Chepko-Sade BD, Halpin ZT (eds) Mammalian dispersal patterns: the effects of social structure on population genetics. University of Chicago Press, Chicago, pp 3–24
Skalski JR, Ryding KE, Millspaugh JJ (2005) Wildlife demography: analysis of sex, age and count data. Elsevier Academic, Burlington
Sokal RR, Rohlf FJ (2012) Biometry, 4th edn. W.H. Freeman and Company, New York
Spinage CA (1972) African ungulate tables. Ecology 53(4):645–652
Thogerson CM, Brady CM, Howard RD, Mason GJ, Pajor EA, Vicino GA, Garner JP (2013) Winning the genetic lottery: biasing birth sex ratio results in more grandchildren. PLoS ONE 8(7):e67,867
Trapletti A, Hornik K (2015) tseries: time series analysis and computational finance. http://CRAN.R-project.org/package=tseries, R package version 0.10-34
Trivers RL, Willard DE (1973) Natural selection of parental ability to vary the sex ratio of offspring. Science 179:90–91
Valenzuela N (2004) Temperature–dependent sex determination. In: Deeming DC (ed) Reptilian incubation: environment and behaviour. Nottingham University Press, Nottingham, pp 211–227
Valenzuela N (2008) Sexual development and the evolution of sex determination. Sexual Development 2(2):64–72
Zar JH (1984) Biological statistics, 2nd edn. Prentice-Hall, Englewood Cliffs

Chapter 6
Ecological Models

Abstract The exponential and logistic population growth models, which are used in ecology to describe population dynamics, are presented. These models can be applied to census data obtained from studbooks. Data from the European blesbok studbook are used to illustrate fitting these models. The method of fitting and testing the exponential growth model as a linear regression model on logarithmic transformed census counts is described. The fitted model and observed census data of the blesbok are compared, and the regression coefficient, which represents the intrinsic rate of increase (r), is tested. The example shows that a significant fit of the exponential population growth model is not necessarily describing the latest trend. This is demonstrated by fitting the logistic growth model using nonlinear least square regression on the blesbok data. The χ^2 goodness of fit test is applied to fitted and observed data in the blesbok example. Interpretation of r in the logistic growth model is discussed. Adaptation of model data to retrieve realistic values for initial population size and start date of the logistic model in R is explored. The advantages and limitations of the "simple" exponential and logistic growth models, and other available methods to analyse time series, are discussed.

6.1 Introduction

Chapters 4 and 5 described population trends in terms of counts (absolute) and rates (relative). These descriptive trends are important in order to understand the basic dynamics of the population. However, it is preferable to describe population trends in terms of models that can be statistically tested and, therefore, can also have predictive value (see also Chap. 11).

The general approach is to find a model and fit a curve. Different models can be used, varying from "simple" linear and exponential curves, multi–degree polynomials (Sokal and Rohlf 2012) to non–linear and time–series regression models as implemented in statistical software such as *R* (R Core Team 2015). Choosing what "complexity" of models to be applied depends on the observations, e.g. patterns of growth and decline, sample size, and on the goal of the study (Bart et al. 1998).

It is worthwhile first to explore some basic ecological models that describe population growth. The *exponential (or geometric)* and *logistic (or sigmoid)* models

F.P.G. Princée, *Exploring Studbooks for Wildlife Management and Conservation*, Topics in Biodiversity and Conservation 17, DOI 10.1007/978-3-319-50032-4_6

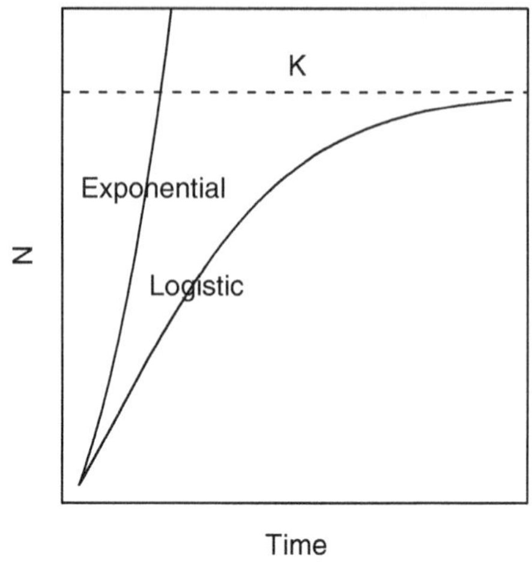

Fig. 6.1 Exponential and logistic growth curves. K = carrying capacity

will be discussed in the following sections. Figure 6.1 illustrates the shape of populations that develop according to these models.

The use of ecological models, especially basic ones, on studbook populations may invoke scepticism, especially since populations are often maintained in the rather safe and controlled environments of zoological gardens. Demographic studies in the 1920s, however, showed that even the development of various human populations followed one of the exponential or logistic models (Pearl 1927). It is important to note that these models do not provide an explanation of why a population developed in such a way. The European studbook for the blesbok will be used, in line with the previous chapters, as a main example in this chapter.

6.2 Population Growth Models

Studbook populations are not expected to be limited by resources such as food and water, or to be exposed to predation. This "unlimited" environment is the condition for exponential growth (see Krebs 1994). However, availability of space, i.e. enclosures and/or semi–free range areas, is a realistic constraint on continuous growth. Space can be considered as the major factor that contributes to the *carrying capacity (*symbol: *K)* of studbook populations. Logistic growth occurs in environments with a carrying capacity. The above suggests that both models may apply during different periods of time in the history of studbook populations.

6.2.1 Exponential Growth

The exponential growth model assumes no limitation in resources such as food, water, nesting sites, territories, etc. The discrete version (e.g. annual growth) of exponential growth is named *geometric growth*.

The exponential model is based on the intrinsic rate of increase (symbol: r):

$$N_t = N_0 e^{rt} \tag{6.1}$$

where t is the time since start, N_t is the population size at time t, N_0 is the population size at $t = 0$ and $e = 2.718282$ (base of natural logarithms) (Krebs 1994).

Equation 6.1 can be rewritten in the form of natural logarithms[1] (Caughley 1977):

$$\ln(N_t) = \ln(N_0) + rt \tag{6.2}$$

This equation represents the linear regression form $y = a + bx$, where slope b is the intrinsic rate of increase (r). Therefore, linear regression on log transformed census counts is the statistical method of choice to test whether actual population growth follows the exponential growth model (Caughley 1977; Sinclair et al. 2006).

Figure 6.2a presents the natural logarithms of the total blesbok census data ($\ln N$) and the linear regression which represents the exponential growth model (Eq. 6.2).

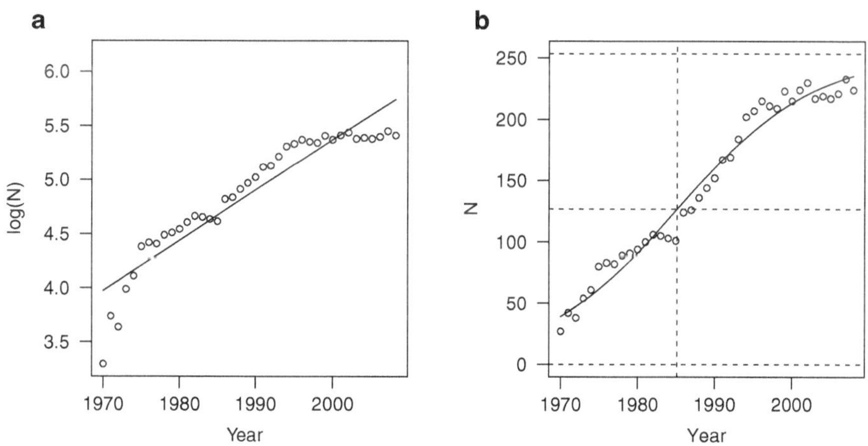

Fig. 6.2 (**a**) Exponential growth model in the blesbok population during the period 1970–2007. Population size N is transformed to the natural logarithm. The *solid line* is the linear regression (see text). (**b**) Logistic growth model. The *solid line* refers to the best fit logistic model. The *upper dashed line* refers to a carrying capacity (K) of 254. The *middle horizontal dashed line* is half K and the *vertical dashed line* is the year in which population size reaches half K (see text)

[1]The symbol for natural logarithm is either ln or $\log_e$.

The intrinsic rate of increase (r) in the exponential model is 0.047. This is lower than the geometric mean of $r = 0.055$ for the period 1970–2007 (calculated from values for λ in Table 4.1). Although the linear fit is significant ($R_{adj}{}^2 = 0.8739$, $F = 264.3, p < 0.001$), the exponential model does not fit the data around the first and last censuses very well.

6.2.2 *Logistic Growth*

The logistic growth model was originally described by Verhulst in 1838 (mentioned in Pearl 1927). This model assumes a carrying capacity that is determined by availability of resources and results in a reduction of the rate of increase proportionate to the increase in population size. The equation for the logistic model can be written as:

$$N_t = \frac{K}{1 + e^{(A-rt)}} \tag{6.3}$$

where K is the carrying capacity, A is a constant, $e = 2.718282$ (base of natural logarithms), r is the intrinsic rate of increase at $t = 0$, and t are the time units (Krebs 1994).

The S–shaped curve of the logistic model explains the alternative name *sigmoid* growth model (see Fig. 6.1). Equation 6.3 is considered the simplest density–dependent growth model as it assumes a linear decline in the intrinsic rate of increase r with increasing population size (N).

The *nonlinear least squares* method can be used to fit a logistic model (Eq. 6.3) onto the total census data (Fox and Weisberg 2010). This method determines the best fit for different values of K, A and r, for which initial values need to be provided. The statistics program R provides functions to calculate the initial values and to determine the best fit for a logistic growth model (Fox and Weisberg 2010; R Core Team 2015).

Figure 6.2b presents total census data for the blesbok and a logistic growth model (solid line) based on the best fit ($K = 253.5$, $A = -223.2$, $r = 0.1124$). However, the "best fit" does not necessarily mean that this model significantly resembles the census data. The *chi–square (χ^2) goodness of fit* test (Sokal and Rohlf 2012) can be used to test the null hypothesis that the census data and the fitted model are from the same distribution.

The χ^2 value of the chi–square test on the blesbok census data and the best fitted logistic model is 19.74. The probability of $\chi^2\, p_{[19.74,38]} = 0.994$ is larger than the significance level $\alpha = 0.05$. Therefore, the hypothesis (H_0), that census data and logistic model are from the same distribution is *not* rejected. This means that the pattern in census data can be explained by a logistic model (with $K = 253.5$, $A = -223.2$, $r = 0.1124$).

The intrinsic rate of increase at $t = 0$ in the logistic model refers to the maximum possible rate (symbol: r_m) that a population, with stable age distribution (see Chap. 7), can reach in its environment when food is abundant and there are no predators, pathogens and competitors (Caughley 1977; Sibly and Hone 2002). The population size N_0 can be calculated from Eq. 6.3 using the estimated values for K, absolute value of A ($|A|$) and r. Note that the R function SSlogis (Fox and Weisberg 2010) uses the logistic growth model with exponent $-(A + rt)$. This results in a negative value for A.

The largest (annual) increase in size occurs when the population has reached half the carrying capacity ($\approx$127). The year that half K is reached can be estimated as $|A|/r = 223.2/0.1124 \approx 1986$.

The value of 3.0e-95 for N_0 does not make sense in a biological context. This value is the result of using the time–unit "century years", which means that t_0 refers to 0 AD. This means that the best fit for the mathematical model that describes logistic growth during the period 1970–2007, is based on a population that started 1,970 years earlier. A more appropriate approach is to index census years to the year when the first historical studbook data with at least a male and female were recorded, in this case 1956 (1 male and 2 females).

Nonlinear regression for different values of t_0 shows that the value for A changes but that values for K and r do not. The results of the χ^2 test also do not change, i.e. census data and logistic model do not differ. For example, the value for A increased from -223.2 to -3.28 when $t_0 = 1956$. This results in a value for N_0 of 9.21 (or 9 individuals in the real world.)

The logistic model describes development of the blesbok studbook population during the period 1970–2007 as logistic growth in a population that started with 9 individuals in 1956, with an initial (maximum) intrinsic rate of increase (r_m) of 0.1124, and a carrying capacity (K) of 254 individuals.

The model is not a perfect fit as the studbook population in reality started with 3 individuals in 1956. However, the chi–square test indicates that the model is a close fit to the real census data. This means that the model can be used in population projections (see Chap. 11).

The estimated carrying capacity, as estimated with the logistic model, is not necessarily related to the available space at studbook locations holding blesbok. However, the reduced population growth since the mid 1990s (Fig. 6.2b) indicates that population size is controlled.

6.3 Remarks

The exponential and logistic models described in the previous section have the advantage that meaningful parameters such as growth rates (r and r_m) are used. However, these models can only be applied to populations with a growth trend in one direction, increasing or decreasing. Ecology is obviously not limited to the simplest logistic model. However, applying more complex ecological models that anticipate predation, species competition or environmental fluctuations is not appropriate for studbook populations, but is used for viability models such as VORTEX (Lacy and Pollak 2014) .

Linear and curvilinear regression are statistical techniques that seem to be appropriate, at first sight, for empirically describing trends in population size. However, there is an important pitfall as the census points are not independent, i.e. the number of individuals in 1 year also depends on the number in the previous year (McDonald 2013). Trends in time series, such as census counts, can be analysed with *Generalised Least Squares (GLS)* in combination with *Auto–Regressive (AR)* models that handle errors due to correlations between data (see e.g. Skalski et al. 2005; Zuur et al. 2009).

References

Bart J, Fligner MA, Notz WI (1998) Sampling and statistical methods for behavioral ecologists. Cambridge University Press, Cambridge

Caughley G (1977) Analyses of vertebrate populations. The Blackburn Press, Caldwell

Fox J, Weisberg S (2010) Nonlinear regression and nonlinear least squares in R: an appendix to an R companion to applied regression, 2nd edn. https://socserv.socsci.mcmaster.ca/jfox/Books/Companion/appendix.html

Krebs CJ (1994) Ecology: the experimental analysis of distribution and abundance, 4th edn. Addison–Wesley, San Francisco

Lacy RC, Pollak JP (2014) VORTEX: a stochastic simulation of the extinction process, version 10.0. Technical report, Chicago Zoological Society, Brookfield

McDonald JH (2013) Handbook of biological statistics, 3rd edn. Sparky House Publishing, Baltimore

Pearl R (1927) The growth of populations. The Q Rev Biol 2(4):532–548

R Core Team (2015) R: a language and environment for statistical computing. R Foundation for Statistical Computing, Vienna. https://www.R-project.org/

Sibly RM, Hone J (2002) Population growth rate and its determinants: an overview. Philos Trans R Soc B 357:1153–1170

Sinclair ARE, Fryxell JM, Caughley G (2006) Wildlife ecology, conservation, and management, 2nd edn. Blackwell Publishing, Malden

Skalski JR, Ryding KE, Millspaugh JJ (2005) Wildlife demography: analysis of sex, age and count data. Elsevier Academic, Burlington

Sokal RR, Rohlf FJ (2012) Biometry, 4th edn. W.H. Freeman and Company, New York

Zuur AF, Ieno EN, Walker NJ, Saveliev AA, Smith GM (2009) Mixed effects models and extensions in ecology with R. Statistics for biology and health. Springer, New York

Chapter 7
Age, Mortality and Fecundity

Abstract Studbooks provide the data to describe life history in terms of age–specific mortality and fecundity tables. Life tables are preferably constructed from following individuals born in the same season (cohort) until all have died. Since studbooks on endangered species are relatively small, virtual cohorts (individuals born over a period of years) are used. Age classes usually have a width of 1 year, but this can be adapted to the biology of a species. Age distributions of living males and females in the studbook population are constructed. The shapes of age pyramids, as described by Bodenheimer, can provide, with some caution, an idea about population dynamics. Construction of mortality tables with mortality rates and associated measures are illustrated with studbook data on snow leopards. The differences between mortality and mortality rate, and between survivorship and survival rate are explained. The differences between mortality tables constructed from recorded deaths in the cohort (uncensored data) and data that are left truncated (staggered entry) and right censored (lost to follow up or living at end of study) are explained. The use of the Mantel–Haenszel (logrank) statistic in comparative survival studies is presented and illustrated by testing differences in survivorship between sexes. Construction of age–specific fecundity rates based on offspring of the same sex as the parents is presented and illustrated with real data. The topics of applying prorating to censored life table data and handling unknown sex and unknown parents are discussed.

7.1 Introduction

Census provides data on numbers of living individuals at given dates (see Chap. 4). However, studbooks also contain data on dates of birth and death. These data provide more detail on the status of a studbook population than "just" the number of living individuals (per sex). This detail means that, for example, an age–distribution can be constructed from studbook data at any census date. The age–distribution will be discussed in Sect. 7.4.

Detailed information on the life history of individuals allows the construction of *life (history) tables* that describe mortality and fecundity at different ages. In the strict sense, a life table refers to mortality (Krebs 1994). Methods to construct these tables were already used by actuaries in the nineteenth century (e.g. to

F.P.G. Princée, *Exploring Studbooks for Wildlife Management and Conservation*,
Topics in Biodiversity and Conservation 17, DOI 10.1007/978-3-319-50032-4_7

calculate annuities and insurance premiums) (Gompertz 1825), before being applied to ecology (Deevey 1947; Pearl and Miner 1935). Both the strict definition of life tables and life tables in which mortality and fecundity are combined, are used in ecological literature (e.g. Begon et al. 2006; Caughley 1977; Krebs 1994).

Studies (and computer software) that involve population projections include fecundity data in life tables (e.g. Ballou et al. 2011; Caswell 2001; Foose and Ballou 1988). These life tables are named *extended life tables* in this book, to avoid confusion in terminology, and are discussed in Chap. 9.

The International studbook of the snow leopard (Blomqvist 2004) is used as the main example in this chapter. This allows the comparison of results of life–table analyses with results from natural history analyses as presented in Chap. 3.

7.2 Cohort

Mortality and fecundity rates are estimated from death and birth events that have occurred in a *cohort*. Within the context of ecology and population biology, a cohort generally refers to a group of individuals of the same sex, who were born in the same breeding season. The cohort is followed over time until the last individual has died.

Since all individuals in a cohort were born in the same birth season, it can be assumed that they have been exposed to similar conditions during their life (see Sect. 3.2.4). However, cohort sizes in studbook populations are in general too small to provide valid results (see Sect. 8.4). For example, the number of annual births, i.e. cohort size, in snow leopards during the period 1983–2001 ranges from 19 to 50 female cubs. These cohort sizes are too small for life history analyses, as will be discussed in Chap. 8.

Studbooks do not necessarily include data which encompass the entire life–span from birth until the last death in a cohort. For example, a studbook on Asian elephants needs to include a sufficiently large cohort of individuals which were born at least 77 years ago.[1]

These limitations mean that cohorts cannot easily be studied in small (studbook) populations. Therefore, an imaginary cohort (Caughley 1977) that combines births (real cohorts) over a period of time is created. The basic imaginary cohort for snow leopards, as used in this chapter, refers to individuals that were born during the years 1983–2002. This period covers a large part of the observed life–span in this species (see Table 3.1).

Each individual is followed from birth until it dies or until the end of the study period (2002), whichever comes first. This means that individuals born in 2002 are followed, depending on day of birth and census date, for at most 12 months. Imaginary cohorts can include individuals that were born before the study period. This will be discussed in Sect. 7.5.3.

[1]This is the observed age of a living Asian elephant in the North American zoo population.

7.3 Age Class

Using "the same breeding season" to define a cohort does not imply that individuals are exactly the same age. The breeding season in birth–flow reproduction systems generally refers to a calendar year or a year between consecutive censuses. Birth–pulse systems can encompass several weeks (see Chap. 3). This means that individuals of a cohort are grouped into an age class.

The class width used in studbook analysis is generally 1 year (e.g. Faust et al. 2012b; Pollak et al. 2005; Scobie and Bingaman Lackey 1997–2012). However, this width can be too large, especially for short–lived species in which individuals reach sexual maturity and reproduce after a few months or even less. Larger class widths may be more appropriate for long–lived species. For example, a class width of 5 years is commonly used in human demography (Shryock et al. 1980). Such class widths could also apply to great apes and elephants. The PMx (Ballou et al. 2011) and PML (Princée 2014) software facilitate flexible class widths.

The initial (first) age class is either numbered 0 or 1. Class 0 is generally used in life tables (e.g. Caughley 1977; Krebs 1994; Skalski et al. 2005), whereas class 1 is more commonly used in population projections based on matrices (e.g. Caswell 2001). This difference may seen trivial, but it can lead to confusion when life table data are used in matrices (see Chap. 11). Demographic management of captive populations has traditionally used the class 0 numbering scheme (e.g. Ballou et al. 2010; Foose and Ballou 1988). The symbol ω is often used for the oldest age class (Keyfitz and Caswell 2005).

7.4 Age Distribution

The first editions of the EAZA/EEP Yearbook included graphic representations of age structure, *age pyramids*, of EEP and ESB populations. The shape of age pyramids provides an overview of population status at a glance. Three major schematic shapes that were presented by Bodenheimer (1938) are often used to interpret age pyramids:

- Pyramid shape with younger age classes that are larger in size than older classes; associated with population growth.
- Bell or pillar shape with age classes more or less similar in size, the oldest age classes can be smaller; associated with a stable population.
- Urn or beehive shape with younger age classes that are smaller in size than older age classes; associated with population decline.

Figure 7.1 presents the age pyramid for the snow leopard population on 1 January 1983. The triangular shape with its broad base (the result of births in 1982), suggests that the studbook population was increasing in that period. However, one needs to

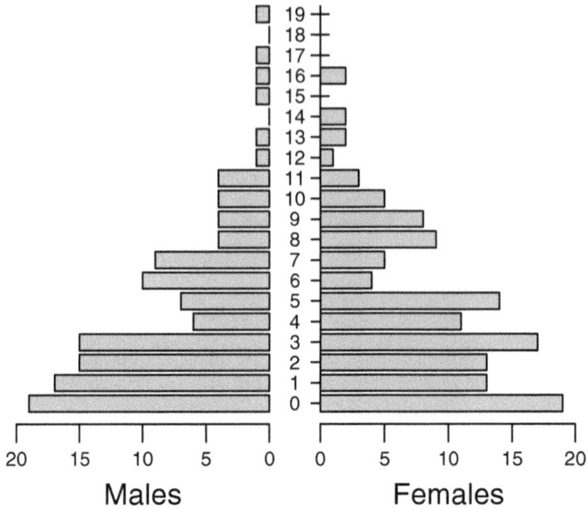

Fig. 7.1 Age distribution of captive snow leopards as on 1 January 1983

be cautious – hence the wording "suggests" – in drawing immediate conclusions from age pyramid shapes (Caughley 1977; Eberhardt 1988).

The interpretation of age pyramid shapes and population growth by Bodenheimer (1938) refers to urban human populations, and does not necessarily apply to other species (Caughley 1977). Furthermore, the age structure can be in a phase of converging to a stable distribution. This means that the shape does not represent the dynamics, but is just a "snapshot" in time (see Chaps. 9 and 11).

It is important to interpret age pyramids in the context of census data. Figure 7.2 shows census data for the period 1983–2002 and age distributions at the start (as in Fig. 7.1) and the end of this census period. The triangular shape of 1983 does indeed represent the characteristics of a growing population. The pillar shape of the 2002 age distribution reflects the stationary size of the previous 4 years.

7.5 Mortality Table

Mortality patterns in a population, such as age–specific mortality rates, are generally presented in table format. Caughley (1977) presents several techniques, depending on the type of data available, for constructing mortality tables. The basic data required are the number of living individuals at the start of each age class (symbol: N_x) – commonly referred to as *individuals at risk* – and the *number of deaths* that occurred within that age class (symbol: D_x).

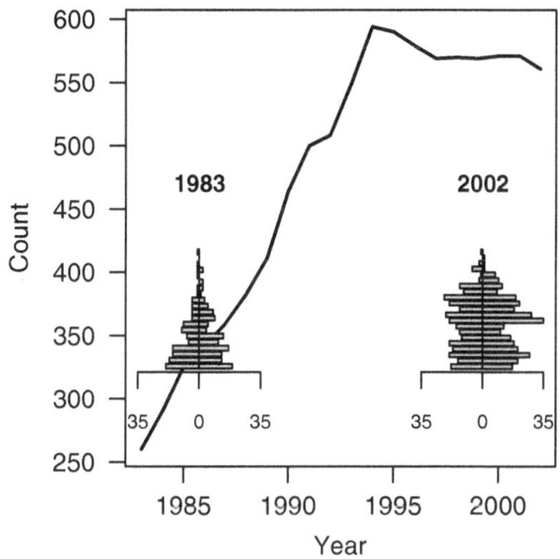

Fig. 7.2 Censuses of the snow leopard population during the years 1983–2002, and age distributions at 1 January for the years 1983 and 2002 (as insets)

7.5.1 *Death Records*

Chapter 3 presented histograms with frequency counts of ages at death. These histograms are important for exploring population data, e.g. to detect patterns and to provide a "feeling" for available sample sizes in relation to statistics. The frequency counts are actually mortality tables, as the data are treated as if all the individuals that died are members of an imaginary cohort.

The D_x values are the frequency counts for age intervals as presented in life-span analysis (Chap. 3); and the cohort size (N_0) is the total number of deaths. The number of individuals at risk of death (N_x) for subsequent age–classes is calculated from the cohort size N_0 minus the total number of deaths that occurred in previous age–classes:

$$N_x = N_0 - \sum_{y=0}^{\omega} D_y \qquad (\textit{for } x > 0) \tag{7.1}$$

where D_y is the number of deaths in age class y; ω is the oldest age class.

7.5.1.1 Mortality Rate

The mortality pattern in a population is generally expressed in terms of *age–specific mortality rates* (symbol: q_x). This measure is defined as the proportion of individuals alive at the start of age class x that died before the end of that age class:

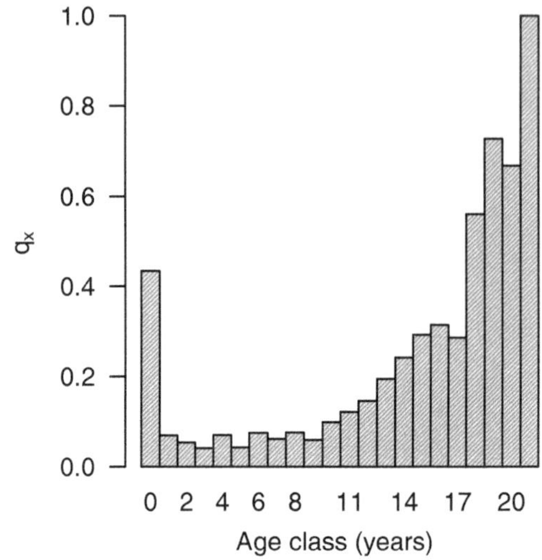

Fig. 7.3 Age–specific mortality rates (q_x) based on death records in captive male snow leopards in the years 1983–2002

$$q_x = \frac{D_x}{N_x} \tag{7.2}$$

where N_x is the number of living individuals at the start of age class x, and D_x is the number of individuals that died before the end of age class x.

Mortality rates reflect the risks of dying during different stages of life. Figure 7.3 shows age specific mortality rates (q_x) in male snow leopards based on deaths that occurred during the years 1983–2002.[2] The mortality rate is high in the class 0 (0–1 year); drops and is more or less stable until class 10; and then increases steeply until the last individual died in class 21. This pattern resembles the "U" shape, and has been observed in various species of large mammal (e.g. Caughley 1966, 1977). In terms of life–stages the pattern describes a high juvenile mortality, low (sub–)adult mortality and high mortality at the end of physical life.

Since numbers of individuals at risk of dying (N_x) decline per age class (Eq. 7.1), number of deaths (D_x) will have high impact on the reliability of mortality rates (q_x) in subsequent classes. For example, the single individual at risk in class 21 died during that class. Consequently, the mortality rate is 1.0 (see Table 7.1).

7.5.2 Mortality Measures

Mortality patterns are not only expressed in terms of mortality rates (q_x). A variety of mortality and survival measures, which are interrelated (Caughley 1977), can be

[2]The year 2003 is excluded as the update of the version used in this book is 17 March 2003.

Table 7.1 Mortality table for male snow leopards based on death records in the years 1983–2002

Age class	N_x	D_x	d_x	q_x	p_x	l_x	L_x	e_x
0	519[a]	228	0.439	0.439	0.561	1.000	0.780	6.188
1	291	21	0.040	0.072	0.928	0.561	0.540	9.644
2	270	14	0.027	0.052	0.948	0.520	0.507	9.356
3	256	11	0.021	0.043	0.957	0.493	0.483	8.840
4	245	15	0.029	0.061	0.939	0.472	0.458	8.214
5	230	9	0.017	0.039	0.961	0.443	0.434	7.717
6	221	15	0.029	0.068	0.932	0.426	0.411	7.011
7	206	13	0.025	0.063	0.937	0.397	0.384	6.485
8	193	13	0.025	0.067	0.933	0.372	0.359	5.889
9	180	10	0.019	0.056	0.944	0.347	0.337	5.278
10	170	17	0.033	0.010	0.900	0.328	0.311	4.559
11	153	19	0.037	0.124	0.876	0.295	0.276	4.010
12	134	19	0.037	0.142	0.855	0.258	0.240	3.507
13	115	21	0.040	0.183	0.817	0.222	0.201	3.004
14	94	23	0.044	0.245	0.755	0.181	0.159	2.564
15	71	21	0.040	0.296	0.704	0.137	0.117	2.232
16	50	16	0.031	0.314	0.680	0.096	0.081	1.960
17	34	10	0.019	0.286	0.704	0.066	0.056	1.647
18	24	13	0.025	0.560	0.458	0.046	0.034	1.125
19	11	8	0.015	0.727	0.273	0.021	0.013	0.864
20	3	2	0.004	0.667	0.333	0.006	0.004	0.833
21	1	1	0.002	1.000	0.000	0.002	0.001	0.500

N_x individual at risk, D_x number of deaths, d_x mortality, q_x mortality rate, p_x survival rate, l_x survivorship, L_x midpoint survivorship, e_x life expectancy

[a] $N_0 = \sum_{x=0}^{\omega} D_x$

derived from q_x. The most common measures are presented in Table 7.1, and will be discussed in this section. Each of these interrelated measures provides a different way of interpreting mortality and survival patterns. The scope of a study dictates which measures are the most appropriate to be presented in a mortality table.

7.5.2.1 Mortality

The measure *mortality* (symbol: d_x) has been briefly discussed in Chap. 3. It refers to the probability that a new–born dies within age class x. Mortality can be calculated as the proportion of deaths in age class x of the total number of deaths:

$$d_x = \frac{D_x}{\sum_{y=0}^{\omega} D_y} \tag{7.3}$$

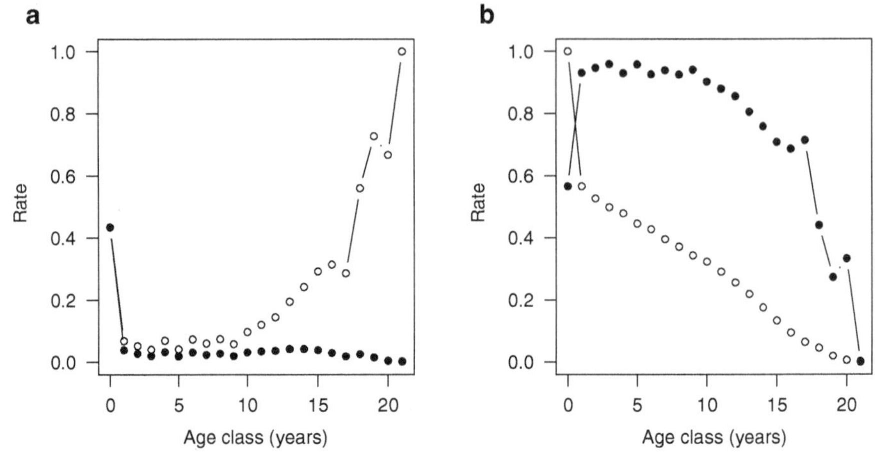

Fig. 7.4 (**a**) Mortality d_x (•) and mortality rates q_x (∘) in captive male snow leopards. (**b**) Survival rate p_x (•) and survivorship l_x (∘)

where D_x and D_y are the number of deaths in age classes x and y, respectively; ω is the oldest age class. The total number of deaths equals the cohort size (N_0) for tables that are based on death records.

Equation 7.3 is not valid for life–tables that are based on censored data (see Sect. 7.5.3). Therefore, the general method of calculating d_x from survivorship rates (l_x) needs to be applied:

$$d_x = l_x - l_{x+1} \tag{7.4}$$

The terms *mortality* and *mortality rate* can give rise to confusion. A way to remember the difference is that d_x is the combined probability that a new–born *survives* up to the start of age class x *and* dies in that age class. The mortality rate (q_x) is the probability that an individual which *has* survived up to the start of class x, dies in that age class. Figure 7.4a shows differences between d_x and q_x for male snow leopards. The main difference between the two measures is that d_x will decline as fewer and fewer individuals survive to the older age class, while q_x will increase to one as finally all individuals will die. See also columns d_x and q_x in Table 7.1.

7.5.2.2 Survival Rate

The *survival rate* (symbol: p_x) is the proportion of individuals alive at the start of age class x that are still alive at the end of that age class:

$$p_x = 1 - q_x \tag{7.5}$$

7.5.2.3 Survivorship

Survivorship (symbol: l_x) is the probability at birth of surviving to the exact age x (which is the start of age–class x). The survivorship at birth (l_0), which is the start of age–class 0, is by definition 1. The survivorship for subsequent age classes ($x > 0$) is calculated as the product ($\prod$) of survival rates (p_x):

$$l_x = \prod_{y=0}^{x-1} p_x \qquad (\textit{for } x > 0) \tag{7.6}$$

The terms survival rate (p_x) and survivorship (l_x) can lead, similarly to mortality (rate), to confusion. Figure 7.4b shows the difference between both measures. Since l_x is the proportion of individuals that have survived until a given age class, this measure will consequently decline towards zero (which it will reach *after* the oldest age class). The survival rate refers to the chance of surviving to the next age class, and will be zero for the oldest age class. See columns p_x and l_x in Table 7.1.

7.5.2.4 Midpoint Survivorship

The *midpoint survivorship* (symbol: L_x) is the probability at birth of surviving within age class x. The average of l_x and l_{x+1} is the most basic method of calculating L_x (Caughley 1977):

$$L_x = (l_x + l_{x+1})/2 \qquad (\textit{for } x < \omega) \tag{7.7a}$$

$$L_\omega = l_\omega/2 \tag{7.7b}$$

where ω is the oldest age class.

Midpoint survivorship is used to estimate the number of individuals that are at risk for reproduction in birth–flow breeding systems.

7.5.2.5 Life Expectancy

The mean life expectancy for an individual that is alive in a given age class (symbol: e_x) can be calculated from survivorship and midpoint survivorship (Shryock et al. 1980):

$$e_x = \frac{\sum_{y=x}^{\omega} L_y}{l_x} \tag{7.8}$$

where L_y and l_x are midpoint survivorship and survivorship, respectively; ω is the oldest age class.

Note that life–expectancy of a new–born snow leopard (class 0) is lower than that of a 1–year old (Table 7.1). The reason is that the risk for a newborn to die in the first year is higher than in the second year. For example, mortality rates in age classes 0 and 1 in male snow leopards are 0.432 and 0.072, respectively (Table 7.1).

7.5.3 Censored Data

Mortality schedules constructed from ages at death that occurred during the study period, do not provide the full "picture" of mortality in a (studbook) population. Individuals that are alive at the end of the study period, and those that emigrated or are "lost to follow–up" during the study period, also contribute to information on survival. These individuals have been "at risk" for each class from birth either to the date when they emigrated (or were lost) or until the end of the study period. This type of incomplete observation is called *right censored* (Kaplan and Meier 1958).

Analyses of ages at death show that husbandry of snow leopards has improved tremendously – measured in terms of increased life–expectancy – since this species arrived in zoological gardens at the end of the nineteenth century (see Chap. 3). Knowledge of natural history of this species was limited in the early days, and consequently husbandry was poor. This means that mortality tables need to be constructed from an imaginary cohort that properly reflects current trends.

The length of study periods, especially in long–lived species, can be too short to provide sufficient data on deaths in the older age classes. A solution to this problem is to include individuals that were born before the start of the study period. These individuals are "at risk" for age classes from the date they entered the study period. This procedure is called *left truncation* or *staggered entry* (Kaplan and Meier 1958; Pollock et al. 1989). Left truncation also applies to immigrants i.e. the individuals are only included for the age classes after they entered the studied population.

Table 7.2 presents a schematic to illustrate left truncated and right censored data in mortality tables based on different life–histories of individuals A to F. The study period in this hypothetical example is from 1 January 2000 to 31 December 2004, and the class–width is 1 year.

Table 7.2 Left truncated and right censored life–table data. See text for explanation

Individual	Truncated	2000	2001	2002	2003	2004	Censored
A		⟹	⟹	⟹	⟹ †		
B				⟹	⟹	⟹	⟹ †
C			⟹	⟹	⟹ ?		
D	(1997) ⟹	⟹	⟹ †				
E	(1999) ⟹	⟹	⟹	⟹	⟹	⟹	⟹
F			(2001)	⟹	⟹ †		
G			(2001)	⟹	⟹ ?		

† = death; ? = lost to follow–up

The following list provides information on how each of these individuals is "treated":

A Born in year 2000 and dies in year 2003. This individual is included as being at risk of dying in age classes 0–3; and counted as a death in age class 3.
B Born in 2002 and dies after the end of 2004. This individual is included as being at risk in age classes 0–3. Since this individual is alive at the end of the time window, it is right censored.
C Born in 2001 and "lost" in 2004. This individual is included as being at risk in age classes 0–2. It is right censored at class 3.
D Born in 1997; arrives in the population in 2000 and dies in 2001. The period 1997–1999 is not considered (left truncated). This individual is included as being at risk in age classes 3 and 4, and counted as a death in class 4.
E Born in 1999 and alive after the end of 2004. This individual is included as being at risk in age classes 1–5. This individual is both left truncated and right censored.
F Born in 2001 outside the population (for example wild–born); arrived in 2002 and died in 2003. This individual is included as being at risk in age classes 2 and 3, and counted as a death in class 3. It is left truncated.
G Born in 2001 outside the population; arrived in 2002 and lost in 2003. This individual is included as being at risk in age classes 2 and 3. It is both left truncated and right censored.

Studbook analyses software such as SPARKS (Scobie and Bingaman Lackey 1997–2012), PopLink (Faust et al. 2012a) and PML (Princée 2014) implement the "rules" as described above to construct mortality tables (and fecundity tables, see Sect. 7.6).

Censored mortality data are left truncated and/or right censored, while data that only include deaths are considered as *uncensored* data.

The method of constructing imaginary cohorts differs between mortality tables based on uncensored and censored data. The cohort based on uncensored data consists of individuals that died during the study period (see Sect. 7.5.1). The cohort size declines according to deaths (D_x) that have occurred in previous age classes (see Eq. 7.3). Note that the death data that are used in this section are not truly uncensored as these data are left truncated, i.e. individuals that were born before the study period are included in the calculations (individuals A, D and F in Table 7.2).

It is difficult to define a single cohort when staggered entries are included. Individuals that were born before the "date window", and/or arrived from another population after the start of the study period, are added as being at risk for the appropriate age class(es). The numbers at risk (N_x) for a given age class x can be larger than the number for a previous class e.g. when individuals of age x were imported in the population (individuals D and E in Table 7.2). This means that Eq. 7.3 is not valid.

Table 7.3 shows mortality rates (q_x) and survivorship (l_x) which are based on left truncated and right censored data in snow leopards during the period 1983–2002. A

Table 7.3 Mortality table for male and female snow leopards based on left truncated and right censored data combined for the years 1983–2002

	Males				Females			
Age class	N_x	D_x	q_x	l_x	N_x	D_x	q_x	l_x
0	702	228	0.325	1.000	725	229	0.316	1.000
1	466	21	0.045	0.675	474	9	0.019	0.684
2	438	14	0.032	0.645	466	7	0.015	0.671
3	419	11	0.026	0.624	454	16	0.035	0.661
4	395	17	0.043	0.608	421	8	0.019	0.638
5	365	10	0.027	0.582	406	11	0.027	0.626
6	348	14	0.040	0.566	380	13	0.034	0.609
7	333	13	0.039	0.543	356	11	0.031	0.588
8	311	13	0.042	0.522	345	24	0.070	0.570
9	285	10	0.035	0.500	317	16	0.050	0.530
10	260	17	0.065	0.482	271	13	0.048	0.503
11	226	19	0.084	0.451	234	21	0.090	0.479
12	194	18	0.093	0.375	197	13	0.066	0.436
13	160	19	0.119	0.330	162	21	0.130	0.407
14	121	23	0.190	0.267	128	18	0.141	0.355
15	89	21	0.236	0.204	103	26	0.252	0.305
16	56	16	0.286	0.146	67	22	0.328	0.228
17	38	9	0.237	0.129	38	12	0.316	0.153
18	28	13	0.464	0.111	21	11	0.524	0.105
19	12	8	0.667	0.060	9	6	0.667	0.050
20	3	2	0.667	0.020	2	0	0.000	0.017
21	1	0	0.000	0.007	2	2	1.000	0.017

N_x individual at risk, D_x number of deaths, q_x mortality rate, l_x survivorship

number of differences between censored (Table 7.3) and uncensored data (Table 7.1) can be observed. Since individuals that were alive at the end of the study period are included, the numbers of individuals at risk per age class in censored data are larger than in uncensored data.

The mortality rate (q_x) in the last age class of a censored mortality table can be lower than 1, as individuals in that class can still be alive by the end of the period of interest. For example, a single male individual in class 21 (Table 7.3) did not die before the end of 2002.[3]

Tables 7.1 and 7.3 indicate that mortality patterns in censored and uncensored data in male snow leopards do differ. For example, mortality (rates) in age class 0 differs significantly between censored and uncensored data ($\chi^2 = 16.46, p < 0.001, df = 1$). Survivorship ($l_x$) curves are useful to visualise differences between

[3]This male is the same individual that holds the longevity record – as on 31 December 2002 – for captive snow leopards (see Chap. 3).

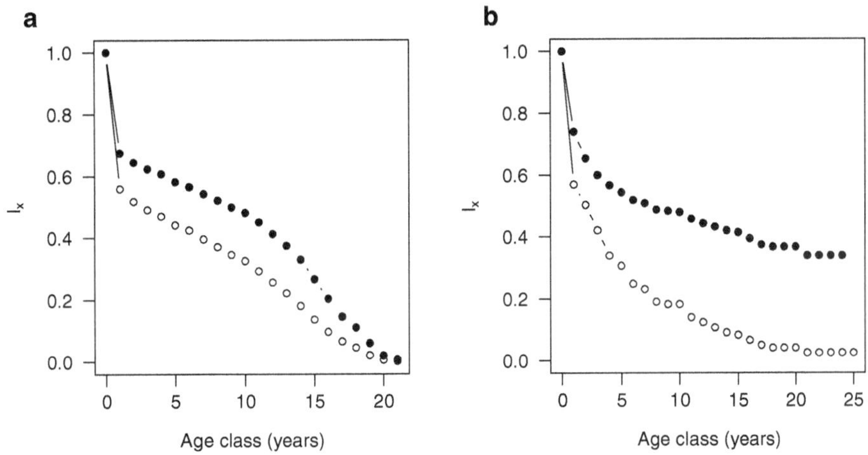

Fig. 7.5 (**a**) Survivorship (l_x) in male snow leopards based on censored (•) and uncensored (○) data from 1983 to 2002. (**b**) Survivorship (l_x) in female red crowned cranes in the European studbook population from 1984 to 2007 based on censored (•) and uncensored (○) data

mortality patterns. Figure 7.5a presents male survivorship for censored (closed dots) and uncensored (open dots) data in snow leopards.

Survivorship based on censored data is higher than that based on death data only. However, differences diminish in the oldest age classes i.e. $\geq$ 20 (see Fig. 7.5a). Since the study period of 20 years[4] is close to longevity in snow leopards (see Chap. 3), differences between both data-sets in these classes are not expected. However, one needs to be cautious in drawing conclusions as sample sizes in the older age classes are very small (see Tables 7.1 and 7.3).

Differences between censored and uncensored data are likely to occur in studbooks of long–lived species, such as the red crowned crane. Longevity in the European studbook population is $\approx$43 years. Figure 7.5b shows differences between survivorship curves for censored (closed dots) and uncensored (open dots) data in female red crowned cranes for the period 1984–2007. Analysis of uncensored data shows that most females have died by the age of 24 ($l_{24} = 0.025$), whereas analysis involving censored data shows that 50 % are still alive at that age.

The red crowned crane example illustrates the importance of using similar types of data (and methods) to compare survivorship, and other measures of mortality, between populations (see also discussions on longevity of elephants in captivity (Mar et al. 2012; Wiese and Willis 2004)).

[4] 1 January 1983–31 December 2002.

7.5.4 Logrank Statistic

The two–way contingency table (Sokal and Rohlf 2012)[5] in the previous section showed that mortality in age class 0 in censored and uncensored male snow leopard data is not the same. However differences, whether subtle or extreme, can occur in any age class. The *logrank statistic*, also known as *Mantel–Haenszel test* (Mantel 1963), can be used to test for differences in survivorship curves between groups that are left truncated and right censored (Pollock et al. 1989). This test is generally used to compare survivorship (l_x) curves in Kaplan–Meier survival analysis (see Chap. 10). However, the logrank test can also be applied to data that are grouped in discrete intervals such as age classes (Morton and Stutchbury 2000; Zhang and Sun 2010). Data on individuals at risk (N_x) and number of deaths (D_x) per age class are used in this test.

The logrank statistic between two groups is calculated as:

$$\chi^2 = \left(\frac{O1 + E1}{E1}\right)^2 + \left(\frac{O2 + E2}{E2}\right)^2 \tag{7.9}$$

where $O1$ and $O2$ are the total number of *observed* deaths in all age classes of groups 1 and 2, respectively; $E1$ and $E2$ are the *expected* totals in these groups. The degree of freedom for this χ^2 statistic is 1 (Peto and Peto 1972). The null hypothesis (H_0) considers the data in both groups to originate from the same distribution, and is rejected when $p_{[\chi^2,1]} < \alpha$.

The total expected number of deaths for group 2 ($E2$) is calculated as:

$$E2 = \sum_{x=0}^{\omega} \frac{D_x}{N_x} N_{2x} \tag{7.10}$$

where D_x and N_x are the combined totals of observed deaths and individuals at risk in age class x in both groups, respectively; ω is the oldest age class; N_2x is the number of deaths in age class x of group 2. The value for $E1$ is calculated as $D_{total} - E2$, where D_{total} is the total number of observed deaths in both groups (Bewick et al. 2004).

The logrank test has been applied to the censored and uncensored data sets of male snow leopards (Tables 7.1 and 7.3). The null hypothesis that survivorship between these groups does not differ is rejected ($\chi^2 = 29.39$, $p < 0.001$, $df = 1$). This result is expected from the significant differences in survivorship between both groups at age class 0 (see previous section). The use of survivorship rates based on right censored and left truncated data are preferred above those based on uncensored, i.e. deaths–only, data (Wiese and Willis 2004).

[5]The choice for Fisher's exact test or the χ^2 test depends on sample sizes.

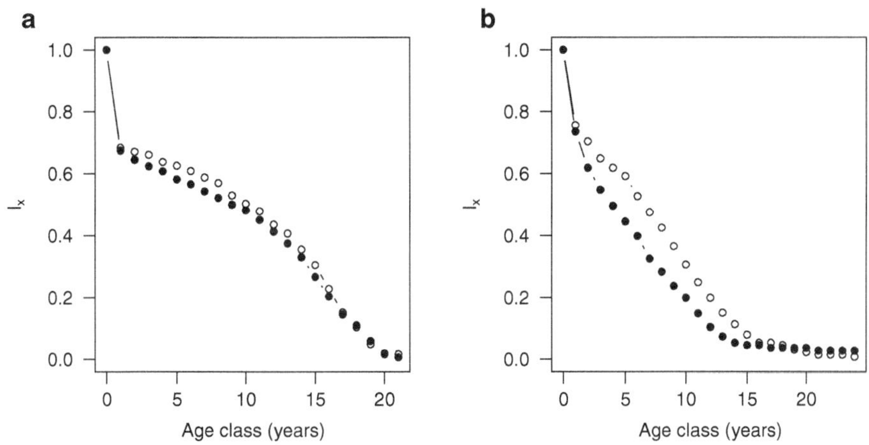

Fig. 7.6 (**a**) Survivorship (l_x) in male (•) and female (○) snow leopards from 1983 to 2002. (**b**) Survivorship (l_x) in male (•) and female (○) blesbok from 1970–2008

7.5.5 Comparative Survival Studies

The logrank statistic is important for comparative survival studies, such as differences between sexes, historical differences in husbandry, environmental conditions (e.g. climate) or differences between inbred and non–inbred groups (see Chap. 13).

Figure 7.6a presents survivorship in male (closed dots) and female (open dots) snow leopards between 1983 and 2002 (data from Table 7.3). Survivorship of males and females does not differ significantly ($\chi^2 = 0.78$, $p = 0.376$, $df = 1$). However, survivorship of male and female blesbok does differ, as would be expected from Fig. 7.6b ($\chi^2 = 26.32$, $p < 0.001$, $df = 1$). This illustrates that graphical presentations of data in comparative studies are helpful in the interpretation of statistical tests.

The logrank test on censored data can be considered as an alternative to the Wilcoxon test, which was discussed in Chap. 3, to test for differences in frequency (count) distributions between ages at death.

No one will think of pseudoreplication in relation to life tables of wild populations. Missing data on parentages and siblings and/or pre–breeding census make it virtually impossible to determine whether the litter effect occurs in juvenile mortality, for example. Studbooks can provide these data and, therefore, enable the testing of litter effects and potential impact on statistical analyses.

Chapter 19 presents a simulation study of the impacts of litter, maternal and spatial effects on sampling variances in neonatal mortality. Chapter 10 presents the *Cox proportional hazard regression model* (Cox 1972) that can take pseudoreplication in survival data into account.

7.6 Fecundity Table

Chapter 3 presents reproductive lifespan analysis, primarily used to determine first and last age at reproduction (see Table 3.4). This data–set can also be used to construct frequency distributions of reproductive ages (see Fig. 3.7a, b). However, it provides only partial information about *fecundity* in the population, as only individuals that have reproduced are included.

Fecundity tables that include animals at risk of reproducing (N_x) per age class and the number of offspring born to parents in that age class (symbol: B_x) provide a better estimate of age–related reproductive success than the life–span analysis. The *fecundity rate* (symbol: m_x) is the mean number of births (B) produced in the group of living individuals (N_x) at the start of each age class x during that age class (see Eq. 7.11).

$$m_x = \frac{B_x}{N_x} \tag{7.11}$$

where B_x is the total number of offspring.

The fecundity rate is not a probability, unlike the mortality rate (q_x), as litter size and number of litters produced by an individual within an age class can result in $B_x > N_x$.

Fecundity tables are generally constructed separately for males and females. Only offspring of the same sex as the parent group are included in B_x. This method seems to be an "inheritance" from wildlife management, where often only the female population and thus female offspring – as reproductive unit – is studied. Fecundity tables that include all offspring are more appropriate to reflect fecundity in the case of unequal sex–ratio at birth. However, such tables cannot be used in standard future projections that are based on Leslie matrices (Leslie 1945) (see Chap. 11). Therefore, fecundity rates that are based on offspring of the same sex as the parent group are used in this section.

Table 7.4 presents fecundity tables for male and female snow leopards. Fecundity rates (m_x) are presented for the sexes separately in Fig. 7.7a, b. Reproductive lifespan is more or less the same in both sexes i.e. age classes 2–15/16. However, the patterns do differ: The highest fecundity rates in females are in classes 4–7, those in males are in classes 3–11.

Caution is needed in interpreting fecundity rates of studbook populations, purely in terms of biological features. For example, management measures to control population size can involve delays in first reproduction and/or limiting the total number of litters per individual. Furthermore, genetic management may temporarily emphasise breeding of genetically important individuals who are near the end of the reproductive lifespan.

Table 7.4 Fecundity table for male and female snow leopards based on left truncated and right censored data from 1983 to 2002

	Males				Females			
Age class	N_x	$N_{bred,x}$	B_x	m_x	N_x	$N_{bred,x}$	B_x	m_x
0	702	0	0	0	725	0	0	0
1	466	0	0	0	479	0	0	0
2	438	3	6	0.014	466	4	5	0.011
3	419	26	39	0.093	454	26	36	0.080
4	395	36	58	0.147	421	42	66	0.158
5	365	30	42	0.115	406	47	62	0.153
6	348	31	55	0.158	380	39	62	0.164
7	333	32	51	0.154	356	35	57	0.161
8	311	27	41	0.132	345	30	39	0.113
9	285	19	28	0.099	317	21	28	0.089
10	260	30	45	0.176	271	19	32	0.118
11	226	21	28	0.124	234	21	28	0.120
12	194	10	16	0.083	197	17	18	0.092
13	160	13	17	0.107	162	6	9	0.056
14	121	5	5	0.041	128	6	6	0.047
15	89	1	1	0.011	103	1	1	0.010
16	56	1	1	0.018	67	0	0	0
17	38	0	0	0	38	0	0	0
18	28	0	0	0	21	0	0	0
19	12	0	0	0	9	0	0	0
20	3	0	0	0	2	0	0	0
21	1	0	0	0	2	0	0	0

N_x individuals at risk, $N_{bred,x}$ individuals that bred, B_x number of births of the same sex as the parent, m_x fecundity rate

7.7 Prorating

The previous sections considered individuals to be at risk of dying or of reproducing from beginning to end of an age class. This method is not necessarily correct when applied to censored data in *studbook views* (studies) that involve date span and/or are limited to geographic populations. Arrivals of individuals that are born outside the studbook view, do not need to coincide with the start of an age class. Likewise, individuals can leave the population before the end of a class.

In both cases, individuals are only partly at risk for these age classes while in the studbook view. This can result in underestimation of mortality and fecundity rates. Proportional distribution of time at risk within an age class (*prorating*) is a method to reduce this bias (Odum and Smith 2001). For example, an individual that enters the studbook view at the age of 8 months would be counted as $\frac{4}{12}$ to be at risk for

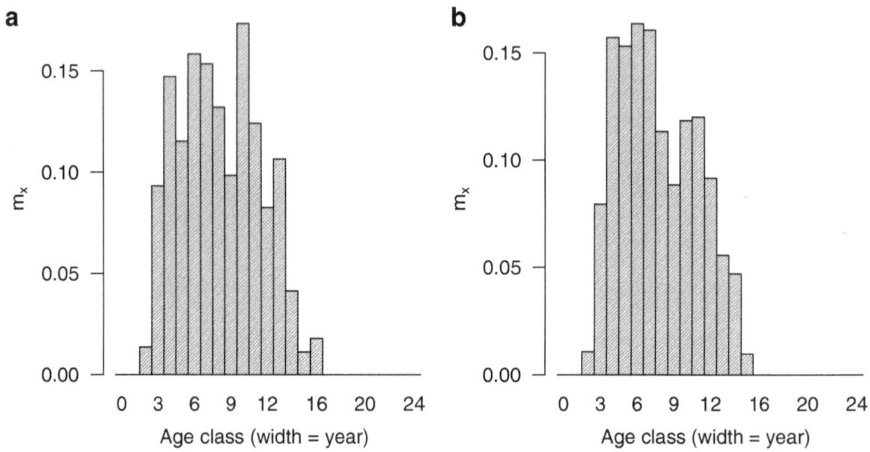

Fig. 7.7 (**a**) Fecundity rates (m_x) in male snow leopards from 1983 to 2002. (**b**) Fecundity rates in female snow leopards from 1983 to 2002

the remaining period of time in age class 0. A newborn individual that leaves the population at the age of 8 months, is effectively at risk for the fraction $\frac{8}{12}$.

Prorating is not applied to the age class in which individuals died, regardless of the time being at risk. Thus, each of the individuals in the examples would be counted as 1 for being at risk if they had died in age class 0.

Since individuals that have died cannot reproduce, they are treated the same as those that have left the study before the end of the age class, i.e. they are prorated for being at risk for reproducing in that class. This approach applies to fecundity in birth–flow species. Individuals in birth–pulse species are effectively no longer at risk for reproduction after the breeding season. This means that prorating should not be applied to individuals that die or leave after the breeding season (Odum and Smith 2001).

The SPARKS and PopLink studbook software apply prorating to age class–based mortality and fecundity tables (Faust et al. 2012b; Odum and Smith 2001). The PML software (Princée 2014) provides combinations of disabling/enabling prorating, left truncation and/or right censoring in life tables.

7.8 Unknown Sex and Unknown Parents

The examples of mortality and fecundity in the previous sections do not consider individuals of unknown sex. Although the subject of "incomplete and missing data" is discussed in more detail in Chap. 18, it is important to be aware of the potential underestimation of mortality and fecundity rates (for sexes separately).

Sex–determination of individuals that have died in the neonatal stage is not always feasible, for various reasons. Unknown sex in mammal studbooks is often associated with perinatal or neonatal death, whereas an individual of known sex has likely survived that stage. For example, 123 snow leopards of unknown sex that were born in the period 1983–2002 all died within the first 3 months, i.e. $q_0 = 1$. Mortality rates estimated for males and females in this age class are 0.325 and 0.316, respectively (see Table 7.3). Including unknown sex, assuming equal distribution over both sexes, increased the mortality rates to 0.378 and 0.369, in males and females, respectively.

Fecundity rates are underestimated when offspring of unknown parents and offspring of unknown sex are not included in the birth counts. Offspring of unknown sex, but with known parents, can be distributed equally over the fecundity age classes of its parents. Offspring with unknown parents can either be distributed equally or according to the m_x ratios (based on known parents) over the fecundity age classes.

7.9 Remarks

Life–table analyses require a higher level of accuracy in dates of birth and death than census analyses. Census analyses only require information on whether an individual was alive at census date. Thus, as long as a birth, death or transfer occurred within census intervals, the individual is counted.

Note that the date estimates in SPARKS use a real date (which is flagged as an estimate) (Scobie and Bingaman Lackey 1997–2012). Protocols have been defined to determine, depending on seasonality or estimated birth and death within the same year, which date estimate is appropriate (Faust et al. 2012b; Thompson et al. 1997). For example, the middle of the year is recommended for species with no distinct breeding season (birth–flow model). These "real" dates do not affect census analyses that are based on "end–of–year counts", but can affect analyses that are based on census dates within a year (see Chap. 18).

Pedigree data are essential to provide parent's ages for calculating age–specific fecundity rates. This means that missing parentages also affect the accuracy of growth rates estimated from fecundity and mortality tables (see Chap. 9). Therefore, census analysis is an important tool to determine population growth in the case that life–table analyses cannot be applied properly.

Mortality tables in studbook analyses are based on methods that are used in wildlife ecology. Mortality data on wild populations are in general collected and/or deduced from annual census data. Studbooks, however, provide greater detail that makes it possible to study mortality – provided that sufficient data per age class are available – in units of months, weeks or even days. The *Kaplan–Meier product limit estimator* (Kaplan and Meier 1958), which finds its origin (partly) in human medical studies, can be applied to "profit" from detailed studbook data. This method will be discussed in Chap. 10.

References

Ballou JD, Lees C, Faust LJ, Long S, Lynch C, Bingaman Lackey L, Foose TJ (2010) Demographic and genetic management of captive populations. In: Kleiman DG, Thompson KV, Kirk Baer C (eds) Wild mammals in captivity: principles and techniques for zoo management, 2nd edn. The University of Chicago Press, Chicago, pp 219–252

Ballou JD, Lacy RC, Pollak JP (2011) PMx: software for demographic and genetic analysis and management of pedigreed populations, version 1.0. Technical report, Chicago Zoological Society, Brookfield

Begon M, Harper JL, Townsend CR (2006) Ecology: from individuals to ecosystems, 4th edn. Blackwell Publishing, Malden

Bewick V, Cheek L, Ball J (2004) Statistics review 12: survival analysis. Crit Care 8(5):389–394

Blomqvist L (2004) International pedigree book for snow leopards (*Uncia uncia*) dataset. Technical report, Helsinki Zoo, Helsinki. Updated to 31 Dec 2003

Bodenheimer FS (1938) Problems of animal ecology. Oxford University Press, London

Caswell H (2001) Matrix populations models: construction, analysis, and interpretation, 2nd edn. Sinnauer Associates, Sunderland

Caughley G (1966) Mortality patterns in mammals. Ecology 47(6):906–918

Caughley G (1977) Analyses of vertebrate populations. The Blackburn Press, Caldwell

Cox DR (1972) Regression models and life-tables. J R Stat Soc Ser B (Method) 34(2):187–220

Deevey ES Jr (1947) Life tables for natural populations of animals. Q Rev Biol 22(4):283–314

Eberhardt LL (1988) Using age structure data from changing populations. J Appl Ecol 25(2):373–378

Faust LJ, Bergstrom YM, Thompson SD, Bier L (2012a) Poplink version 2.4. Lincoln Park Zoo, Chicago

Faust LJ, Bier L, Schowe K, GazLay T (2012b) PopLink version 2.4: user's manual. Lincoln Park Zoo, Chicago

Foose TJ, Ballou JD (1988) Population management: theory and practice. Int Zoo Yearb 27:26–41

Gompertz B (1825) On the nature of the function expressive of the law of human mortality, and on a new mode of determining the value of life contingencies. Philos Trans R Soc Lond 115:513–583

Kaplan EL, Meier P (1958) Nonparametric estimation from incomplete observations. J Am Stat Assoc 53:457–481

Keyfitz N, Caswell H (2005) Applied mathematical demography, 3rd edn. Statistics for Biology and Health. Springer, New York

Krebs CJ (1994) Ecology: the experimental analysis of distribution and abundance, 4th edn. Addison-Wesley, San Francisco

Leslie PH (1945) On the use of matrices in certain population mathematics. Biometrika 33(3):183–212

Mantel N (1963) Chi–square tests with one degree of freedom: extensions of the Mantel–Haenszel procedure. J Am Stat Assoc 58(3):690–700

Mar KU, Lahdenperä M, Lummaa V (2012) Causes and correlates of calf mortality in captive Asian elephants (*Elephas maximus*). PLoS ONE 7(3):32335

Morton ES, Stutchbury BJM (2000) Demography and reproductive success in the dusky antbird, a sedentary tropical passerine. J Field Ornithol 71(3):493–500

Odum RA, Smith BR (2001) The effects of prorating risk in the development of life-tables. Zoo Biol 20:279–291

Pearl R, Miner JR (1935) Experimental studies on the duration of life. XIV. The comparative mortality of certain lower organisms. Q Rev Biol 10(1):60–79

Peto R, Peto J (1972) Asymptotically efficient rank invariant test procedures. J R Stat Soc Ser A (Gen) 135(2):185–207

Pollak JP, Lacy RC, Ballou JD (2005) Population Management 2000, version 1.213. Chicago Zoological Society, Brookfield

Pollock KH, Winterstein SR, Bunck CM, Curtis PD (1989) Survival analyses in telemetry studies: the staggered entry design. J Wildl Manag 53:7–15
Princée FPG (2014) Population Management Library (PML) software and manual, version 1.0. East Rudham. http://www.princee.com
Scobie P, Bingaman Lackey L (1997–2012) Single Population Animal Record Keeping System (SPARKS) versions 1.42–1.66, International Species Information System, Eagan
Shryock HS, Siegel J (1980) U.S. Bureau of the census – the methods and materials of demography, 4th edn. U.S. Government Printing Office, Washington, DC
Skalski JR, Ryding KE, Millspaugh JJ (2005) Wildlife demography: analysis of sex, age and count data. Elsevier Academic, Burlington
Sokal RR, Rohlf FJ (2012) Biometry, 4th edn. W.H. Freeman and Company, New York
Thompson SD, Earnhardt JM, Turner-Erfort G (1997) Guidelines for data entry and maintenance of North American regional studbooks. Lincoln Park Zoo, Chicago
Wiese RJ, Willis K (2004) Calculation of longevity and life expectancy in captive elephants. Zoo Biol 23:365–373
Zhang Z, Sun J (2010) Interval censoring. Stat Method Med Res 19(1):53–70

Chapter 8
Confidence in Life Tables

Abstract Confidence limits of age–specific mortality and fecundity rates are rarely provided. However, this information is important for judging the reliability of life tables, especially those based on small studbooks. The occurrence of a single death within an age class is a Bernoulli process, i.e. an individual is either dead or living at the end of the age class, with the mortality rate as the probability of dying. Simulation experiments using random sampling introduce the effect of sample size on the confidence limits of mortality rates. Variances of mortality rates can be calculated from binomial theory. Fecundity rates are often assumed to follow the Poisson distribution. The variance is estimated accordingly. Variances, confidence limits and coefficients of variation are estimated for both mortality and fecundity in young and older age classes. Bootstrap is applied to measure variance (and associated statistics) in both mortality and fecundity rates in female snow leopards. The results of χ^2 tests show that theoretical and bootstrap variances do not differ for mortality (except for the oldest age class). However, these variances do differ for fecundity, indicating that the Poisson distribution is not the correct assumption. Reliability of life tables can be improved by increasing sample sizes through pooling older age classes and applying wider age classes (at the cost of detail). Smoothing techniques to reduce gaps and peaks in life tables and application of the Gompertz(–Makeham), Weibull and Siler parametric models to fit mortality curves, are illustrated with real studbook data.

8.1 Introduction

Fecundity and mortality rates, particularly in older age classes, are often based on relatively small sample sizes (see Tables 7.4 and 7.3, respectively). One can intuitively judge that rates based on sample sizes of e.g. 3, 5 or 10 do not represent the "true" mortality, whereas those based on sample sizes of 1,000 are more likely to do so. But what about sample sizes in between? Variances and confidence intervals provide more insight about appropriate sample sizes.

This chapter presents methods to calculate variances and confidence intervals based on expected probability distributions of fecundity and mortality. However, real–time data do not necessarily follow these distributions. For example, deaths in litters are not necessarily independent events, e.g. in cases of poor parental care all

F.P.G. Princée, *Exploring Studbooks for Wildlife Management and Conservation*, Topics in Biodiversity and Conservation 17, DOI 10.1007/978-3-319-50032-4_8

litter mates may die. Computer resampling techniques can be applied to estimate variances and confidence intervals without knowledge of the underlying probability distribution. The *bootstrap* method (Efron 1979) is presented in this chapter.

Variances will remain high when sample sizes per age class are small. Pooling of older age classes, wider age classes and smoothing to remove data "peaks" due to random effects, can be applied to improve life tables. These techniques are presented in this chapter. In addition, parametric models that are applied to fit mortality data are presented.

8.2 Tossing a Coin

The occurrence of a single death within an age class is a *Bernoulli* process (a special case of a binomial process) that can be summarised as *an individual either being alive or dead* at the end of that age class. The mortality rate (q_x) can be considered as the probability (risk or chance) of dying, and $1-q_x$ is consequently the probability of survival. It is like tossing a coin (an unfair one whenever $q_x \neq 0.5$).

Assume that the observed neonatal mortality rate is 0.3. One would expect that 30 out of 100 newborns in the next breeding season will die before reaching the age of one year. However, in reality the actual number may differ purely by chance from the expected number. This random process can be illustrated with simulation experiments by sampling randomly from a binomial distribution with the number of newborns as "trials" and the mortality rate as probability (p). Each sample represents a number of newborns that died.

Figure 8.1a presents results of a simulation experiment with 10,000 random samples (repeats). Although the mean mortality is close to 30 individuals, sample values range from 14 to 47. The 2.5 and 97.5 % percentiles, which together are an estimate of the 95 % confidence interval, are 21 and 39, respectively. This means that the expected mortality rate for 100 newborns will likely be in the range 0.21–0.39.

The impact of this random process, named *demographic stochasticity* or *uncertainty* (Shaffer 1981), increases for smaller numbers. Figure 8.1b presents a similar simulation experiment based on 10 newborns (trials) and 10,000 samples. The mean mortality is again close to the expected 3 individuals, but the 95 % confidence interval ranges from 0 to 6 (or a mortality rate from 0 to 0.60).

The confidence interval will narrow around the mean for increasing numbers of individuals. For example, increasing the number of newborns (trials) to 1,000 resulted in 2.5 and 97.5 % percentiles for mortality rates of 0.27 and 0.33, respectively.

These simulation experiments provide an insight into the effects of demographic uncertainty on mortality rates in relation to numbers of individuals at risk. Mortality rates based on small numbers at risk are *not* likely to reflect the "true" mortality. This means that such data are not suitable for use in comparative studies and population projections (see Chap. 11). The next section will discuss variances and confidence intervals for mortality rates in more detail.

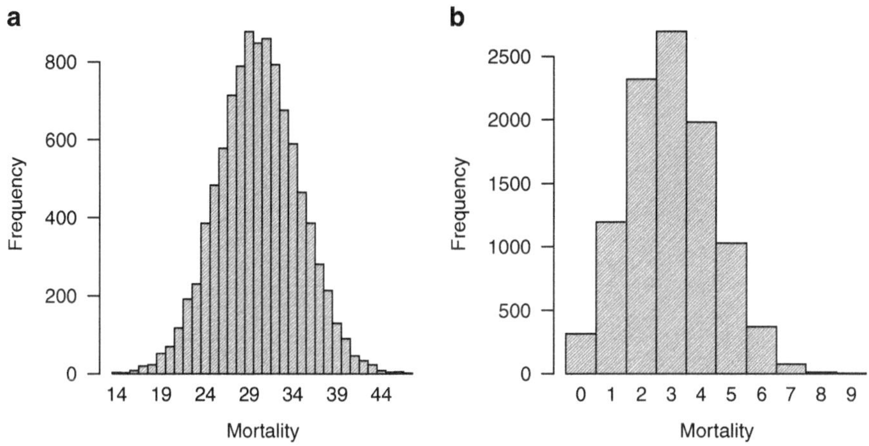

Fig. 8.1 (**a**) Frequency distribution of neonatal mortality in 10,000 random samples from a binomial distribution with 100 trials (newborns) and probability of death (p) is 0.3. (**b**) Frequency distribution based on a binomial distribution with 10 trials

8.2.1 Mortality

Considering mortality as a binomial process means that the variance ($s^2_{q_x}$) of a mortality rate can be calculated as:

$$s^2_{q_x} = \frac{q_x(1 - q_x)}{N_x} \tag{8.1}$$

where q_x is the mortality rate and N_x the number of individuals at risk in age class x (Daw 1974).

The 95 % confidence interval for mortality rates is calculated as $q_x \pm 1.96\sqrt{s^2_{q_x}}$. The confidence limits may need to be adjusted whenever they are outside the 0–1 range of mortality rates. Additionally, the coefficient of variation or "relative variability" (symbol: CV) can be estimated:

$$CV = \frac{s_x}{\bar{x}} \tag{8.2}$$

where $\bar{x}$ is the mean and s_x is the standard deviation of parameter x (Sokal and Rohlf 2012). Since the mortality rate is a mean, $\bar{x}$ can be substituted with q_x.

Coefficients of variation are often presented as percentages (by multiplying with 100). Since CVs are "unit free", they can be used to compare relative variability among age classes and between different life tables.

Table 8.1 presents theoretical variances, 95 % confidence interval, and coefficients of variation of mortality rates in female snow leopards. The coefficient of variation (CV), or "relative variability", provides the information on which to judge

Table 8.1 Variance, 95 % confidence interval and coefficient of variation (as percentage) of mortality rates in youngest and oldest age classes of female snow leopards. Mortality is assumed to be a binomial process

Age class	N_x	D_x	q_x	$s^2_{binomial}$	Confidence interval		CV(%)
					Lower	Upper	
0	725	229	0.3195	0.00030	0.2821	0.3497	5.48
1	479	9	0.0188	0.00004	0.0064	0.0310	33.64
…							
18	21	11	0.5238	0.01188	0.3102	0.7374	20.81
19	9	6	0.6667	0.02469	0.3587	0.9747	23.57

N_x individuals at risk, D_x number of deaths, q_x mortality rate, $s^2_{binomial}$ theoretical variance in mortality rate (binomial distribution), *CV(%)* coefficient of variation in percentages

whether variance is low or (too) high. For example, the variance in class 19 is higher, as expected from samples sizes, than the variance in class 1. However, the opposite trend can be observed in the CV values.

8.2.2 *Fecundity*

The effects of random processes on fecundity rates can be expected to be similar to those discussed for mortality rates in Sect. 8.2. The major difference is that the expected number of births is often assumed to be distributed according to the Poisson model (Akçakaya 2002). The variance of fecundity rates ($s^2_{m_x}$) that follow the Poisson process is calculated as:

$$s^2_{m_x} = \frac{m_x}{N_x} \tag{8.3}$$

where m_x is the fecundity rate and N_x the number of individuals at risk of reproduction in age class x (Keilman and Pham 2000). The coefficient of variation (CV) is estimated by substituting $\bar{x}$ with m_x, in Eq. 8.2.

Table 8.2 presents theoretical variances ($s^2_{Poisson}$), 95 % confidence limits and coefficients of variation (CV) in fecundity rates. Several methods exist to calculate exact confidence limits for the Poisson distribution (Patil and Kulkarni 2012). The "Garwood method" in *R* is used (Fay 2010; R Core Team 2015).

The *CVs* for age classes with low numbers of births are large, e.g. the value for age class 15 is over 100 %! Although a single birth was observed in this age class, the upper confidence limit indicates that $103 \times 0.0323 \approx 3$ female offspring (from 103 females) could have been expected by chance alone.

Table 8.2 Variance, 95 % confidence interval and coefficient of variation (as percentage) of fecundity rates at beginning, middle and end of reproductive age–span of female snow leopards. Fecundity is assumed to be a Poisson process

					Confidence interval		
Age class	N_x	B_x	m_x	$s^2_{Poisson}$	Lower	Upper	CV(%)
2	466	5	0.0108	0.00002	0.0035	0.0250	41.4
3	454	36	0.0795	0.00018	0.0555	0.1098	16.9
...							
6	380	62	0.1635	0.00043	0.1251	0.2092	12.7
7	356	57	0.1605	0.00045	0.1213	0.2074	13.2
...							
14	128	6	0.0470	0.00037	0.0172	0.1020	40.9
15	103	1	0.0097	0.00009	0.0000	0.0323	102.0

N_x individuals at risk, B_x number of births, m_x fecundity rate, $s^2_{Poisson}$ theoretical variance in fecundity rate (Poisson distribution), *CV(%)* coefficient of variation in percentages

The assumption that births are a Poisson process is not necessarily justified (Devenish–Nelson et al. 2013; Kendall and Wittman 2010). Births in species that are *monovular* (producing a single egg during each cycle, e.g. blesbok and California sea lions) is a Bernoulli process: *an individual gives birth or not.* Fecundity in such a species can be modelled, similar to mortality, as a binomial distribution (Caswell 2001).

Modelling births in *polyovular* species (producing multiple offspring during each cycle, e.g. snow leopards and African wild dogs) is more complex than in monovular species, as probabilities of giving birth and of litter size need to be combined. Discrete distributions such as binomial, negative binomial and Poisson are candidates for describing the underlying model (e.g. Devenish–Nelson et al. 2013).

Fecundity rates are low in the younger and older age classes, e.g. Fig 7.7b. This is due to a large proportion of non–breeders (zero litters) and possibly smaller litter sizes (in polyovular species). The number of zero litters may exceed the number expected from the binomial or Poisson distribution. This means that the variances in the younger and older age classes may be better described by *zero–inflated* distribution models which account for large numbers of zero litters, e.g. zero–inflated Poisson model (e.g. Devenish–Nelson et al. 2013). These distribution models are certainly more appropriate for describing male fecundity in species that live in harem structures in which most non–harem holder males do not breed (e.g. those in bachelor herds).

The above illustrates the difficulties in making assumptions regarding the underlying probability distribution of fecundity in polyovular species and males of harem species. The following section introduces a method that is free of assumptions regarding probability distributions.

8.3 Resampling

Variances and confidence limits in mortality and, particularly, in fecundity rates cannot be estimated when the underlying probability distributions are unknown. Since only one studbook population (in time and space) exists it is not possible to estimate variances from a sample of life tables.

The studbook population itself can be considered as the best available "image" of the unknown imaginary population from which it was drawn. Statistical characteristics, such as mean and variance, can then be estimated by resampling the observed (studbook) data. This approach is the backbone of resampling techniques such as the *jackknife* (Quenouille 1949; Tukey 1958) and the *bootstrap* (Efron 1979).

The jackknife technique systematically leaves out each observation once from the original data set of size N (thus, each sample size is $N - 1$) and recalculates the value(s). The removed observation is included again before the next sample is taken. The mean and other statistics are estimated from the total $N - 1$ samples.

The bootstrap technique draws random samples (replicates or bootstraps) (symbol: B) with replacement from the original data. Each sample is of the same size as the original. This means that observations can occur more than once or not at all in a sample. The mean and other statistics are estimated from the B samples.

Increased access to computers has resulted in more common use of these techniques, particularly bootstrap, in biological research (Crowley 1992; Davidson and Hinkly 1997). It is not surprising that resampling techniques have also been applied to assess uncertainties in mortality and fecundity rates (Caswell 2001; Efron 1981; Meyer et al. 1986).

Bootstrap and jackknife resulted in similar variances of growth rates that were estimated from life tables (Meyer et al. 1986). However, the jackknife performs poorly in estimating variances of sample medians (Efron and Tibshirani 1986). This means that this resampling technique is not suitable for analysis of median lifespan. Therefore, the bootstrap technique has been implemented in the life table analyses of the PML software (Princée 2014).

Individuals are sampled randomly with replacement from the (imaginary) studbook cohort which can include censored data (see Chap. 7). The life history of each individual in the cohort is "followed" to determine in which age classes it was at risk, in which age class it died, or in which age class it reproduced (fecundity and mortality analyses are carried out separately in PML). This process is repeated in each bootstrap. Mean and other statistics are estimated for each age class from the bootstrap samples.

The number of bootstrap replicates that are required depends on the statistic (see Davidson and Hinkly 1997). This number of $B = 1{,}000$ is considered as a rough minimum to estimate bootstrap confidence intervals (Efron and Tibshirani 1986). The number of replicates is adjusted when an exact test (with significance level α) is to be carried out: the result of $\alpha(B + 1)$ must be an integer (Davidson and MacKinnon 2000). The minimum of 1,000 bootstraps adjusted for $\alpha = 0.01$ results in $B = 999$, which is used in this book.

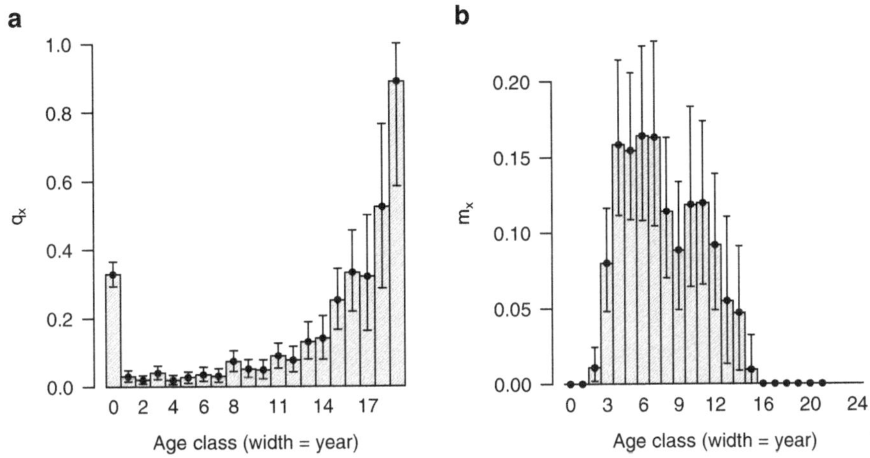

Fig. 8.2 (**a**) Bootstrap results ($B = 999$) of mortality rates (q_x) in female snow leopards from 1983–2002. The histogram refers to the mean bootstrap value. The 2.5 and 97.5 % percentiles are indicated with *horizontal lines* per age class. (**b**) Bootstrap results ($B = 999$) of fecundity rates (m_x) in female snow leopards

Figures 8.2a and 8.2b present results of bootstrap on mortality and fecundity rates (left censored and right truncated) in female snow leopards, respectively. The bars refer to the mean rates over 999 replicates (B). The lower and upper horizontal bars refer to the 2.5 and 97.5 percentiles, and can be interpreted as a basic (95 %) confidence interval (Davidson and Hinkly 1997).

The patterns in confidence intervals in Figs. 8.2a and 8.2b are similar to those observed in the data presented in Tables 8.1 and 8.2. However, these figures do not indicate whether the bootstrap rates differ from those calculated for the original data, and/or whether variances differ from those calculated from probability distributions.

Table 8.3 combines results from the bootstrap analyses on fecundity and mortality with the data in Tables 8.1 and 8.2. A chi–square test ($\alpha = 0.05$) is applied to test whether the sample (i.e. bootstrap) and theoretical variances (see Eqs. 8.1 and 8.3) are the same (Sokal and Rohlf 2012). A plus sign means that this null hypothesis is *not* rejected, and hence that the bootstrap variance and theoretical variance do *not* differ.

Statistical tests on mortality rates in age classes 0–18 (not all presented in Table 8.3), do not reject the null hypothesis. The rejection for age class 19 is likely to be due to the small sample size (Table 8.1). These results justify the conclusion that female mortality in the studbook population of snow leopards is a Bernoulli process. This means that variances and confidence intervals can be calculated using Eq. 8.1.

The bootstrap fecundity rates ($\bar{m}_x$) and the observed rates merely differ at the third decimal (this level of accuracy may not even be appropriate given the sample sizes for females being at risk). First, these differences are likely to be due to the

Table 8.3 Comparison of bootstrap analysis on mortality and fecundity rates with observed data in female snow leopards. See text for details

Mortality					
Age class	q_x	$s^2_{binomial}$	$\bar{q}_x$	$s^2_{bootstrap}$	χ^2test[a]
0	0.3297	0.00030	0.3286	0.00030	+
1	0.0292	0.00006	0.0297	0.00006	+
...					
18	0.5238	0.01188	0.5238	0.01268	+
19	0.8889	0.01097	0.8883	0.01329	-
Fecundity					
Age class	m_x	$s^2_{Poisson}$	$\bar{m}_x$	$s^2_{bootstrap}$	χ^2 test[a]
2	0.0108	0.00002	0.0108	0.00003	-
3	0.0795	0.00018	0.0799	0.00028	-
...					
6	0.1635	0.00043	0.1641	0.00081	-
7	0.1605	0.00045	0.1631	0.00092	-
...					
14	0.0470	0.00037	0.0471	0.00036	+
15	0.0093	0.00009	0.0093	0.00009	+

q_x mortality rate, $s^2_{binomial}$ binomial variance q_x, $\bar{q}_x$ bootstrap q_x, m_x fecundity rate, $s^2_{Poisson}$ Poisson variance m_x, $\bar{m}_x$ bootstrap m_x

[a] + H_0 not rejected: theoretical and bootstrap variances do not differ

stochastic nature of the bootstrap sampling process. A larger number of sampling repeats can result in mean values that are closer to the observed values. Second, the differences in observed and bootstrap fecundity rates are not that dramatic, especially in the scope of small studbook populations. For example, the 380 females in age class 6 would have produced, with $\bar{m}_x = 0.1641$, 62.38 cubs (which are 62 in real life), instead of the observed 62 (Tables 8.2 and 8.3).

Although differences between observed and bootstrap fecundity rates in female snow leopards seem to be small, theoretical and bootstrap variances for fecundity rates do differ significantly, except for the older age classes with low birth numbers (Table 8.3).

Bootstrap of life history data shows that variances in age–specific mortality in female snow leopards follow a binomial distribution, except for the last age class. However, these variances can differ from the theoretical variances (Eq. 8.1) when pseudoreplication occurs (see Chap. 19).

Fecundity rates do not necessarily follow a Poisson or other known distribution, as illustrated in female snow leopards. The advantage of bootstrap is that it provides approximate confidence intervals (percentiles) regardless of the type of distribution.

8.4 Sample Sizes

The previous section illustrated the effects of demographic stochasticity on age classes with small numbers of individuals at risk. In studbook populations this generally refers to the older age classes (Tables 7.3 and 7.4). Studbook populations that have been established recently can have small sample sizes in all age classes. A solution to the sample size problem is to reorganise the data by "pooling" older age classes and/or increasing class–width.

8.4.1 Pooling

Pooling the data from older age classes in the tail of the distribution into a single age class can be applied to life–tables in order to create a larger sample size. The resulting class is generally indicated with the addition '+'. Table 8.4 shows the effect of pooling age classes 18–21 into the single age class 18+ on mortality rates. The samples sizes of the four classes are small, but combining these data will result in an age class where the mortality rate is 1.0 based on 24 individuals at risk. A class size of around 30 individuals is recommended (Ballou et al. 2010).

Since the data in this example are uncensored, the number of individuals at risk *and* deaths in class 18+ is the sum of deaths in classes 18–21. This relation does not exist in censored mortality data where individuals can be lost to follow–up. Reconstruction of N_x and D_x from observed mortality rates (q_x) for the classes to be pooled is required.

Pooling age classes in fecundity tables from censored data is less straightforward than explained for mortality. First the number of individuals at risk in the pooled age classes needs to be reconstructed from mortality rates as described previously. The next step is to calculate the expected number of births (B_x) from the reconstructed N_x and (observed) fecundity rates (m_x). The actual pooling is carried out on the reconstructed table. The PML studbook software will automatically pool individuals that live longer than the maximum requested age–class (Princée 2014).

Table 8.4 Pooling older mortality classes (male snow leopard data from Table 7.1)

	Observed			Pooled	
Age class	N_x	D_x	q_x	D'_x	q'_x
18/18+	24	13	0.560	24	1.000
19	11	8	0.727		
20	3	2	0.667		
21	1	1	1.000		

N_x individuals at risk, D_x number of deaths, q_x mortality rate, D'_x pooled number of deaths, q'_x pooled mortality rate in class 18+

8.4.2 Class–Width

Studbook life tables are generally based on class–widths of one year, which is common in wildlife management where census tends to be carried out annually (Caughley 1977). Such a width is too wide for short–lived species, especially those that reach sexual maturity and reproduce within the first year. Using one year would result in loss of information on mortality in the neonatal, juvenile and sub–adult life stages (see Chap. 3). In contrast, one year intervals can result in large numbers of – less informative – age classes in long–lived species. For example, human demographers tend to use class–widths of five years (e.g. Shryock et al. 1980).

Recent studbook software includes the option of selecting class–width, varying from day, week, month(s) to years (Ballou et al. 2011; Princée 2014). Class–width should ideally reflect a species' life history. However, the available data in studbooks may not be sufficient to gauge this, as explained in the previous sections. Statistical methods to construct frequency distributions (histograms) can serve as guidelines to selecting appropriate class–widths.

A basic recommendation in statistical literature is to base frequency distribution (histograms) on 12–20 classes (e.g. Sokal and Rohlf 2012). This translates to class–widths of one year for species with a longevity in the range of 12–20 years; six months for species with a longevity of 5–12 years etc. Class–widths of 2–5 years would be appropriate for species like elephants and great apes. However, this recommendation does not consider the total sample size.

The *Sturges algorithm* (Sturges 1926)[1] could be applied to estimate potential class–width (symbol: w):

$$w = \frac{\Omega}{1 + 3.322 * ln(N)} \tag{8.4}$$

where Ω is the (observed) longevity and N the total sample size.

Recommended class–width for mortality and fecundity data on snow leopards (see Tables 7.3 and 7.4) ranges between 0.83 and 1.23 years, for total sample sizes of 100 to 1,000, respectively. Sturges (1926) recommends using "convenient" class–widths such as 1, 2, 5, 10, 20 (units). This means that a class–width of one year seems to be appropriate for species with a longevity of 20 years, such as snow leopards. The Sturges algorithm does not consider the distribution of data and thus does not necessarily reflect low sample sizes at the tail of mortality and fecundity. This means that pooling of older age classes may still be required.

Increasing class–width for left truncated and/or right censored mortality and fecundity tables can be achieved "manually" as outlined for pooling (see Sect. 8.4.1). The new class–width must be a multiplication of the original width with an absolute natural number (i.e. 1,2,3,4 ...etc.).

[1]The Sturges algorithm is implemented as default method in R statistics to estimate bin (class) width.

8.5 Smoothing

Small sample sizes of studbook data not only result in limited or lack of data in older age classes, but in intermediate classes as well. Furthermore, the distribution may show peaks that do not necessarily reflect the "true" mortality or fecundity distribution. Figure 8.3a illustrates gaps and "unexpected" peaks, e.g. class 6, in mortality rates (q_x) in male wolverines. The available data in this studbook can be considered small, i.e. $Nm_0 = 56$, and data of older age classes are pooled in class 15+ ($N_{15+} = 7$).

The "noise", i.e. gaps and peaks, can be removed by applying *smoothing* to the original data–set. There is no "one" smoothing method. Basic methods involve running (moving) averages or medians, and variations thereof, which can often be calculated by hand (e.g. Tukey 1977). More complex methods involve smoothing splines (e.g. Horton and Kleiman 2011).

The SPARKS software implements smoothing that is based on running averages of three, for both mortality and fecundity rates (Scobie and Bingaman Lackey 1997–2012). The PMx software implements a running median of three (Ballou et al. 2011). Neither software program smooths age class 0. Furthermore, SPARKS calculates the rate for age class 1 as $rate_1 = (rate_1 + rate_2)/2$.

SPARKS and PMx extend the life table with an extra age class ($\omega + 1$) in order to calculate the rate at the endpoint. SPARKS uses a rate of 0 for class $\omega + 1$, where PMx uses $rate_{\omega+1} = 3 \times rate_{\omega-2} - 2 \times rate_{\omega-1}$. Smoothing can be repeated recursively, i.e smoothing of smoothed values, in both SPARKS and PMx.

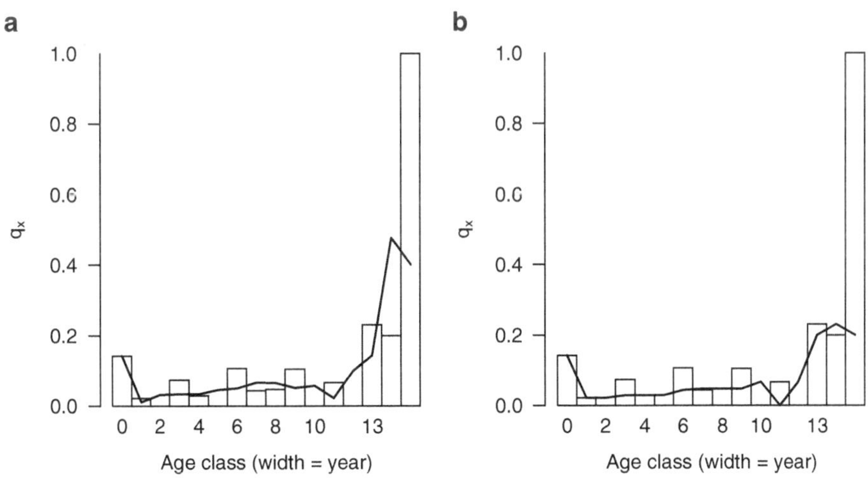

Fig. 8.3 (**a**) Smoothed mortality rates (*line*) for male wolverines using the SPARKS algorithm of running means (no repeats). The histogram presents actual values. (**b**) Smoothing using the PMx algorithm of running medians (no repeats)

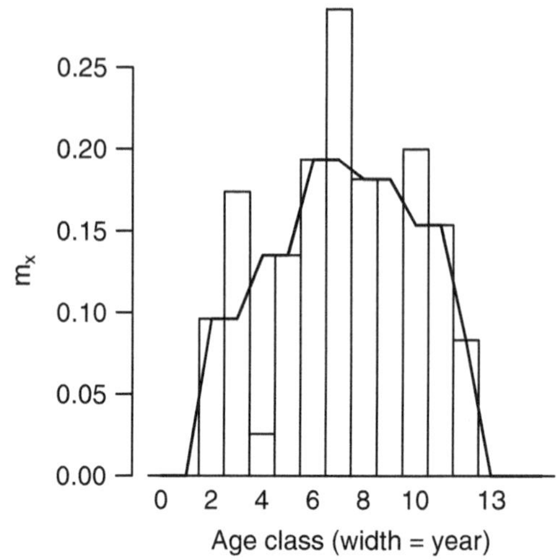

Fig. 8.4 Smoothed fecundity rates (*line*) for female wolverines using the PMx algorithm of running means (no repeats). The histogram presents actual values

Figures 8.3a and 8.3b present smoothing of mortality rates in male wolverines following the SPARKS and PMx methods, respectively. Both smoothing procedures reduce the gaps and peaks in the observed rates. The smoothed values for the last age class (15) are considerably lower than the observed q_{15} for both methods. This can be expected as mortality rates in the previous age classes are also considerably lower i.e. $\approx$0.2. Since the PMx method places more emphasis on the two age classes previous to the last one, the smoothed value for q_{15} is lower than calculated by the SPARKS method.

Smoothing can also be applied to fecundity rates. Figure 8.4 presents observed and smoothed fecundity rates (black line) for female wolverines using the PMx "method" of running medians (Ballou et al. 2011). Smoothing clearly results in reducing the gaps (class 4) and peaks (classes 3, 7 and 10) as observed in the original data.

Smoothing does not assume an underlying model. Its main purpose, as described previously, is to reduce the noise in observed data in order better to visualise potential trends with age. This noise is due to random sampling effects and has, therefore, more impact on small data–sets. For example, observed mortality rates in male snow leopards ($Nm_0 = 702$) show a more fluent distribution than those observed in male wolverines ($Nm_0 = 56$; see Figs. 8.2a and 8.3a, respectively). It is important to evaluate results of smoothing with the original data, especially in relation to sample sizes, and/or species biology (Traylor–Holzer 2011).

8.6 Parametric Models

Smoothing procedures may not be sufficient to "bridge" data gaps between age classes. A further step could be to test whether parametric models used in demography will fit the data. The gap data can then be interpolated from the fitted model.

Actuarial science has been a driving force in developing the first models to describe mortality patterns in human populations. The *Gompertz model* (Gompertz 1825) and the extended *Gompertz–Makeham model* (Makeham 1860) were developed in the 1800s and are still in use today, not only in actuarial science, but in ecology and (bio–)gerontology as well (e.g. Carriere 1992; Kohler et al. 2006; Ricklefs and Scheuerlein 2001).

The Gompertz model, and other mortality models that are discussed in this section, are generally presented as *(instantaneous) hazard rates* (symbol: h_x). Other names for the hazard rate in the context of life tables are *force of mortality* and *instantaneous rate of mortality* (e.g. Klein and Moeschberger 1997; Ricklefs and Scheuerlein 2002).

The instantaneous rate of mortality (h_x) is the proportion of living individuals at age x that die per unit of time. This measure differs from the mortality rate (q_x) which is expressed as the proportion of individuals at risk that die within a fixed period of time i.e. class interval x.

$$h_x = h_0 e^{\gamma x} \tag{8.5}$$

where h_0 is the initial instantaneous rate of mortality experienced in young adults and γ is the exponential rate of increase in mortality rate with age (Gompertz 1825; Ricklefs and Scheuerlein 2002).

The Gompertz distribution is an exponential function, which means that it can only describe increasing mortality rates i.e. in older age classes. The Gompertz–Makeham model (Makeham 1860) is an extension of the Gompertz model that better fits the early mature age classes where mortality rates are more or less constant.

$$h_x = h_0 + a e^{\gamma x} \tag{8.6}$$

where h_0 and a are constants that partition the initial instantaneous rate of mortality; γ is the exponential rate of increase in mortality rate with age (Gompertz 1825; Ricklefs and Scheuerlein 2002).

The parameters h_0, a and γ need to be solved in order to fit the Gompertz(–Makeham) model to observed data. Statistical programs such as the R package *fmsb* (Nakazawa 2015) provide functions to estimate these constants. Moreover, this package conveniently converts h_x values to mortality rates (q_x).

Figure 8.5a presents the Gompertz–Makeham curve for mortality rates (q_x) in male wolverines. It illustrates the limitation of this exponential model, i.e. the initial decrease in mortality cannot be modelled (see Eqs. 8.5 and 8.6).

The Weibull distribution (Weibull 1951) is often preferred in comparative studies of mortality (or survival) as it better fits early mature mortality (Pinder III et al. 1978; Ricklefs and Scheuerlein 2001; Wiese and Willis 2004). The equation is:

$$h_x = h_0 + \alpha x^{\beta} \tag{8.7}$$

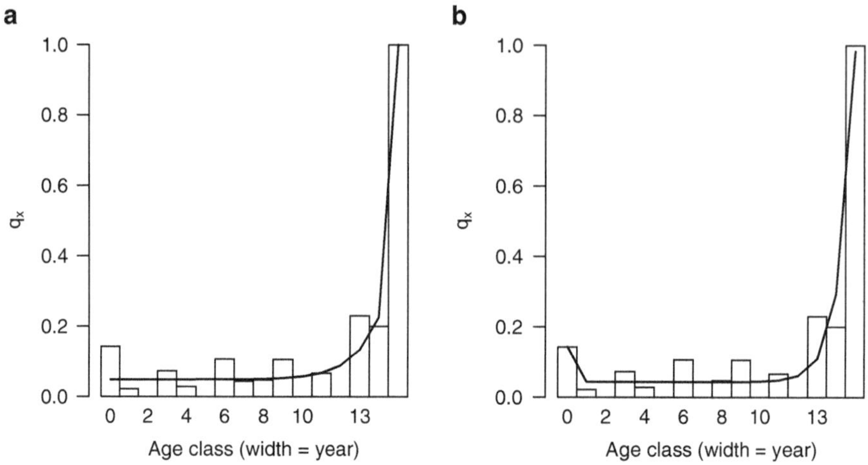

Fig. 8.5 (**a**) Gompertz–Makeham curve for mortality rates in male wolverines. The histogram presents actual values. (**b**) Siler curve.The histogram presents actual values

where h_0 is the initial instantaneous rate of mortality; and αx^{β} is the age dependent instantaneous mortality rate (Ricklefs and Scheuerlein 2002).

The initial mortality in the Weibull model refers, as in the Gompertz(–Makeham) model, to the mature (young adult) stage where mortality rates are more or less constant. Thus, this model also does not cover the juvenile stage. The Weibull model is an important parametric model that is used, for example, in studying ageing in both birds and mammals (Moorad et al. 2012; Ricklefs and Scheuerlein 2001).

The Siler model has been developed to study mortality over the entire lifespan in animal species (Siler 1979). The model is partitioned into immature, mature (young adults) and ageing stages:

$$h_x = h_0 e^{-\gamma_0 x} + h_1 + h_2 e^{\gamma_1 x} \tag{8.8}$$

where h_0, h_1 and h_2 are initial hazard rates for the immature, mature and ageing stages, respectively; and $(-)\gamma_0$ and γ_1 refer to changes in mortality in the immature and ageing stage, respectively.

The partition $h_0 e^{-\gamma_0 x}$ describes the initial decreasing mortality from birth to young adult. Parameter h_1 refers to mortality in the mature stage, which can be compared with h_0 in the Gompertz(–Makeham) and Weibull models. The partition $h_2 e^{\gamma_1 x}$ is similar to the Gompertz model in describing ageing (Eq. 8.5).

Figure 8.5b presents the Siler curve for mortality rates (q_x) in male wolverines. This curve was constructed with the fmsb package. The Siler model results in a better fit to juvenile mortality in this species than the Gompertz–Makeham model does (Fig. 8.5a).

Parametric models for fecundity rates are less well developed than those for mortality rates (Gage 2001). Research in this field mainly focusses on human populations (Peristera and Kostaki 2007; Williamson and Norman 2011). The Brass polynomial and Hadwiger function that are used in human demography to model fecundity have been applied to some mammalian species (Gage 1995, 2001). More studies on the use of these models in different groups of species are required.

References

Akçakaya HR (2002) Estimating the variance of survival rates and fecundities. Anim Conserv 5:333–336

Ballou JD, Lees C, Faust LJ, Long S, Lynch C, Bingaman Lackey L, Foose TJ (2010) Demographic and genetic management of captive populations. In: Kleiman DG, Thompson KV, Kirk Baer C (eds) Wild mammals in captivity: principles and techniques for zoo management, 2nd edn. The University of Chicago Press, Chicago, pp 219–252

Ballou JD, Lacy RC, Pollak JP (2011) PMx: software for demographic and genetic analysis and management of pedigreed populations (version 1.0). Technical report, Chicago Zoological Society, Brookfield

Carriere JF (1992) Parametric models for life tables. Trans Soc Actuar 44:77–99

Caswell H (2001) Matrix populations models: construction, analysis, and interpretation, 2nd edn. Sinnauer Associates, Sunderland

Caughley G (1977) Analyses of vertebrate populations. The Blackburn Press, Caldwell

Crowley PH (1992) Resampling methods for computation-intensive data analysis in ecology and evolution. Ann Rev Ecol Syst 23:405–447

Davidson AC, Hinkly DV (1997) Bootstrap methods and their application. Cambridge series in statistical and probabilistic mathematics. Cambridge University Press, Cambridge

Davidson R, MacKinnon JG (2000) Bootstrap tests: how many bootstraps? Econom Rev 19(1):55–68

Daw RH (1974) A study on the variance of mortality rates. J Inst Actuar 101:415–434

Devenish–Nelson ES, Stephens PA, Harris S, Soulsbury C, Richards SA (2013) Does litter size variation affect models of terrestrial carnivore extinction risk and management? PLoS ONE 8(2):e58060

Efron B (1979) Bootstrap methods: another look at the jackknife. Ann Stat 7(1):1–26

Efron B (1981) Censored data and the bootstrap. J Am Stat Assoc 76:312–319

Efron B, Tibshirani R (1986) Bootstrap methods for standard error, confidence intervals, and other measures of statistical accuracy. Stat Sci 1(1):54–75

Fay MP (2010) Two-sided exact tests and matching confidence intervals for discrete data. R J 2(1):53–58

Gage TB (1995) A robust estimator of the intrinsic rate of increase and generation length. In: Ballou JD, Gilpin M, Foose TJ (eds) Population management for survival and recovery: analytical methods and strategies in small population conservation. Methods and cases in conservation science series. Columbia University Press, New York, pp 9–23

Gage TB (2001) Age-specific fecundity of mammalian populations: a test of three mathematical models. Zoo Biol 20:487–497

Gompertz B (1825) On the nature of the function expressive of the law of human mortality, and on a new mode of determining the value of life contingencies. Philos Trans R Soc Lond 115:513–583

Horton NJ, Kleiman K (2011) Using R for data management and statistical analysis and graphics. CRC Press, Boca Raton

Keilman N, Pham DQ (2000) Predictive intervals for age–specific fertility. Eur J Popul 16:41–66

Kendall BE, Wittman ME (2010) A stochastic model for annual reproductive success. Am Nat 175(4):461–468
Klein JP, Moeschberger ML (1997) Survival analysis: techniques for censored and truncated data, 1st edn. Statistics for biology and health. Springer, New York
Kohler IV, Preston SH, Lackey LB (2006) Comparative mortality levels among selected species of captive animals. Demogr Res 15:413–434
Makeham WM (1860) On the law of mortality and the construction of annuity tables. Assur Mag J Inst Actuar 8(6):301–310
Meyer JS, Ingersoll CG, McDonald LL, Boyce MS (1986) Estimating uncertainty in population growth rates: jackknife versus bootstrap techniques. Ecology 67(5):1156–1166
Moorad JA, Promislow DEL, Flesness N, Miller RA (2012) A comparative assessment of univariate longevity measures using zoological animal records. Aging Cell 11(6):940–948
Nakazawa M (2015) fmsb: functions for medical statistics book with some demographic data. http://CRAN.R-project.org/package=fmsb. R package version 0.5.2
Patil VV, Kulkarni HV (2012) Comparision of confidence intervals for the poisson mean: some new aspects. Revstat Stat J 10(2):211–227
Peristera P, Kostaki A (2007) Modeling fertility in modern populations. Demogr Res 16(6):141–194
Pinder III JE, Wiener JG, Smith MH (1978) The Weibull distribution: a new method of summarizing survivorship data. Ecology 59(1):175–179
Princée FPG (2014) Population Management Library (PML) software and manual version 1.0. East Rudham, United Kingdom, http://www.princee.com
Quenouille MH (1949) Approximate test of correlation in time–series. J R Stat Soc Ser B (Method) 11(2):68–84
R Core Team (2015) R: a language and environment for statistical computing. R Foundation for Statistical Computing, Vienna. https://www.R-project.org/
Ricklefs R, Scheuerlein A (2001) Comparison of aging-related mortality among birds and mammals. Exp Gerontol 36(4–6):845–857
Ricklefs RE, Scheuerlein A (2002) Biological implications of the Weibull and Gompertz models of aging. J Gerontol Ser A Biol Sci Med Sci 57(2):B69–B76
Scobie P, Bingaman Lackey L (1997–2012) Single Population Animal Record Keeping System (SPARKS) Versions 1.42–1.66. International Species Information System, Eagan
Shaffer ML (1981) Minimum population sizes for species conservation. BioScience 31(2):131–134
Shryock HS, Siegel J (1980) U.S. Bureau of the census – the methods and materials of demography, 4th edn. U.S. Government Printing Office, Washington, DC
Siler W (1979) A competing-risk model for animal mortality. Ecology 60(4):750–757
Sokal RR, Rohlf FJ (2012) Biometry, 4th edn. W.H. Freeman and Company, New York
Sturges HA (1926) The choice of class interval. J Am Stat Assoc 21:65–66
Traylor–Holzer K (ed) (2011) PMx users manual, Version 1.0. IUCN SSC Conservation Breeding Specialist Group, Apple Valley, MN
Tukey JW (1958) Bias and confidence in not–quite large samples. (abstract of preliminary report). Ann Math Stat 29(2):614
Tukey JW (1977) Exploratory data analysis. Behavioural science: quantitative methods. Addison-Wesley, Reading
Weibull W (1951) A statistical distribution function of wide applicability. J Appl Mech 18(3):293–297
Wiese RJ, Willis K (2004) Calculation of longevity and life expectancy in captive elephants. Zoo Biol 23:365–373
Williamson L, Norman P (2011) Developing strategies for deriving small population fertility rates. J Popul Res 28(2):129–148

Chapter 9
Extended Life Table

Abstract Life table analysis of studbooks combines mortality and fecundity data in a single table. This table is named in this book "extended life table" to avoid confusion, as life table can also refer to mortality data alone. This chapter shows methods of estimating net reproductive rate (R_0), generation time ($\bar{T}$) and intrinsic rate of increase (r) from age–specific mortality and fecundity rates. The method to calculate Fisher's reproductive value (V_x) is described. This measure refers to the number of expected future offspring (of the same sex) produced by an individual of a given age that is weighted against population growth. The unweighted value is used as a measure of fitness. The age distribution of a population that is growing geometrically with constant mortality and fecundity rates will gradually approach a fixed or stable age distribution. The proportion of individuals in different age classes can be estimated from age–specific survivorship (l_x) and the finite rate of increase (λ). The use of midpoint survivorship (L_x) to estimate the proportional number of births in species that breed all year round (birth–flow) is discussed. The last section of this chapter compares life tables and census data to estimate population growth.

9.1 Introduction

Age–specific fecundity and mortality tables of a cohort are often combined in a single life table (Ballou et al. 2010; Begon et al. 2006). Since life table is also defined as mortality table (Krebs 1994), the combined table in this book is named *extended life table*. The extended life table also includes measures that can be derived from mortality and fecundity rates, such as reproductive value (symbol: V_x) and stable age distribution (symbol: C_x).

Extended life tables contain a large number of parameters by combining mortality, fecundity and derived measures. Therefore, shortened versions that only include the essential mortality and fecundity parameters required for study are often used. Table 9.1 presents a shortened version limited to survivorship (symbol: l_x), fecundity rate (m_x) and derived measures. The next sections will discuss the use of information in this table.

F.P.G. Princée, *Exploring Studbooks for Wildlife Management and Conservation*, Topics in Biodiversity and Conservation 17, DOI 10.1007/978-3-319-50032-4_9

Table 9.1 Extended life table for female snow leopards based on left truncated and right censored data from 1983 to 2002 (inclusive)

Age class	l_x	m_x	$l_x m_x$	V_x	VF_x	C_x
0	1.000	0.000	0.000	1.000	0.760	0.078
1	0.684	0.000	0.000	1.409	1.112	0.055
2	0.671	0.011	0.007	1.384	1.133	0.056
3	0.661	0.080	0.053	1.344	1.139	0.058
4	0.638	0.157	0.100	1.263	1.098	0.058
5	0.626	0.153	0.096	1.087	0.960	0.059
6	0.609	0.164	0.100	0.925	0.829	0.060
7	0.588	0.160	0.094	0.760	0.689	0.060
8	0.570	0.113	0.065	0.596	0.545	0.060
9	0.530	0.088	0.047	0.500	0.464	0.058
10	0.503	0.118	0.060	0.418	0.396	0.058
11	0.479	0.120	0.057	0.303	0.292	0.057
12	0.436	0.092	0.040	0.194	0.189	0.054
13	0.408	0.056	0.023	0.106	0.104	0.052
14	0.355	0.047	0.017	0.056	0.055	0.047
15	0.305	0.010	0.003	0.010	0.010	0.042
16	0.228	0.000	0.000	0.000	0.000	0.033
17	0.153	0.000	0.000	0.000	0.000	0.023
18	0.105	0.000	0.000	0.000	0.000	0.016
19	0.050	0.000	0.000	0.000	0.000	0.008
20	0.017	0.000	0.000	0.000	0.000	0.003
21	0.017	0.000	0.000	0.000	0.000	0.003

l_x survivorship, m_x fecundity rate, $l_x m_x$ average number of offspring produced by individuals in age class x, V_x unweighted reproductive value, VF_x Fisher's reproductive value, C_x is the proportion of the population in age class x under a stable age distribution

9.2 Growth Rates

The *net reproductive rate* (symbol: R_0) is the mean number of offspring (of the same parental sex) produced by each cohort member until the end of the cohort (i.e. death of all members) (Krebs 1994). This measure is calculated from survivorship and fecundity rates:

$$R_0 = \sum_{x=0}^{\omega} l_x m_x \tag{9.1}$$

where ω is the oldest age class; l_x and m_x are survivorship and fecundity rate for age class x, respectively.

Values of R_0 are interpreted as:

- $0 < R_0 < 1$: population size decreases.
- $R_0 = 1$: population size does not change.
- $R_0 > 1$: population size increases.

The net reproductive rate projects population size in the next generation, when conditions do not change. Generation refers to *generation time* (symbol: $\bar{T}$), which is defined as the mean age of parents at date of birth of their offspring. This measure is calculated as follows:

$$\bar{T} = \left(\sum_{x=0}^{\omega} x l_x m_x\right) / R_0 \tag{9.2}$$

where ω is the oldest age class; l_x and m_x are survivorship and fecundity rate for age class x, respectively; and R_0 is the net reproductive rate.

The net reproductive rate and generation time for the female snow leopard population during the period 1983–2002 is 0.724 and 8.36 years, respectively. Note that the time unit for generation time refers to the class–width, i.e. 1 year in this example. This means that the female population is expected to decrease by a proportion of $1 - 0.724 = 0.276$ ($\approx 28\,\%$) in ≈ 8 years time.

The intrinsic rate of increase (r) can be estimated from life table data by solving the *Euler–Lotka equation*:

$$\sum_{x=0}^{\omega} e^{-rx} x l_x m_x = 1 \tag{9.3}$$

where ω is the oldest age class; l_x and m_x are survivorship and fecundity rate for class x, respectively.

Studbook software such as PML (Princée 2014) and PMx (Ballou et al. 2011) implement algorithms to solve the Euler–Lotka equation. However, r can also be approximated when $R_0 \approx 1$ (Begon et al. 2006):

$$r \approx \ln(R_0)/\bar{T} \tag{9.4}$$

where R_0 is the net reproductive rate, and $\bar{T}$ is the generation time.

Equations 9.3 and 9.4 both result, as expected from R_0, in an r value of   0.03. The finite rate of increase (λ) is 0.96 (see Eq. 4.3). Note that λ refers to annual growth when the class–width is 1 year.

9.3 Reproductive Value

Management of studbook populations involves transfers (including translocation) of individuals to create new breeding pairs and groups in order to avoid inbreeding and to maintain genetic variation (see Chaps. 13 and 15). Although genetic importance of individuals is the key selection criterion, other criteria such as fecundity and life expectancy play a role as well. For example, individuals that have reached the post–reproductive stage are no longer of interest for breeding. On the other end of the spectrum are the juveniles who have a high risk of dying before reaching sexual maturity (see Chap. 7). From a demographic perspective, selecting individuals in age groups that have high chances of producing offspring (i.e. low mortality and high fecundity) during their remaining life time is preferred.

The *reproductive value* (symbol: *Vx*) is the number of offspring (of the same sex as the parent) that an individual in age class (x) is expected to produce at present and in the future:

$$V_x = \frac{\sum_{y=x}^{\omega} l_y m_y}{l_x} \tag{9.5}$$

where ω is the oldest age class; l_x is survivorship for age class x; l_y and fecundity rate m_y refer to the remaining age classes (Krebs 1994).

Since reproductive value takes survival chances into account when estimating expected offspring, this measure is useful in planning transfers of individuals to create new breeding combinations. The reproductive value for individuals in age class 0 is the total expected number of offspring in their lifetime, which is the net reproductive rate (*R*0) as discussed in the previous section.

Equation 9.5 is a "simpler" version of the original concept by Fisher (1930), in which reproductive value refers to the relative importance ("value") of an individual's contribution to the future generation. *Fisher's reproductive values* (symbol: $V_{F,x}$) are weighted for population growth (r):

$$V_{F,x} = \frac{\sum_{y=x}^{\omega} e^{-ry} l_y m_y}{e^{-rx} l_x} \tag{9.6}$$

where ω is the oldest age class, r is the intrinsic growth rate, l_x is survivorship for age class x, l_y and fecundity rate m_y refer to the remaining age classes. Fisher's reproductive value for age class 0 is 1.0 (Galindo 2008).

Equation 9.5 estimates the reproductive value in terms of absolute numbers of offspring. Therefore this measure is referred to as *absolute reproductive value* in this book. This measure is the same as Fisher's reproductive value when the population is stable in size (i.e. e^{-ry} and e^{-rx} in Eq. 9.6 are 1 when $r = 0$).

Figure 9.1 shows absolute (open dots) and Fisher's reproductive values (closed dots) for female snow leopards based on fecundity and mortality data during the period 1983–2002. Reproductive values do initially increase as a result of reduced mortality after the juvenile stage and increased fecundity. Although fecundity rates are highest in age classes 6 and 7 (see Fig. 7.7b), the highest reproductive value is observed in age class 3.

Since population size is decreasing ($R_0 < 1$) there will be fewer breeders in the future. Therefore, Fisher's reproductive values of individuals in the (living) population are higher than the absolute reproductive values.

Fisher's reproductive values are used to calculate *kinship values* in selecting genetically important individuals as part of population management (Ballou and Lacy 1995). Both types of reproductive values can be considered as measures of

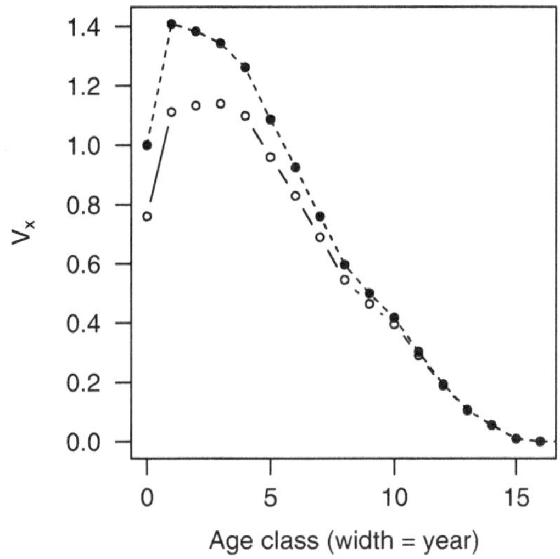

Fig. 9.1 Fisher's (•) and absolute (○) reproductive values for female snow leopards from 1983 to 2002

fitness that can be used in comparative studies between groups/subpopulations. Chapter 13 discusses the use of absolute reproductive values in detecting inbreeding depression.

9.4 Stable Age Distribution

The age distribution of a population that is growing geometrically (discrete version of exponential growth, see Chap. 6) with constant mortality and fecundity rates will gradually approach a fixed or *stable age distribution* (Lotka 1922). The proportion of the population per age class (C_x) can be estimated from the finite rate of increase (λ) and survivorship (l_x) (Mertz 1970):

$$C_x = \frac{\lambda^{-x} l_x}{\sum_{y=0}^{\omega} \lambda^{-y} l_y} \tag{9.7}$$

where ω is the oldest age class; l_x is survivorship for age class x; l_y refers to the remaining age classes.

The stable age distribution for the female snow leopard population, based on 1983–2002 life table data, is presented in Table 9.1. Some 8 % of the individuals in the female population are (female) newborns; the proportions decline per age class.

Figure 9.2 presents the observed female age distribution at post–breeding census and the stable age distribution (both in numbers of animals). The older age classes of the observed age distribution represent a pyramid, indicating population growth in the past. The younger classes more represent the bottom of a bell–shaped

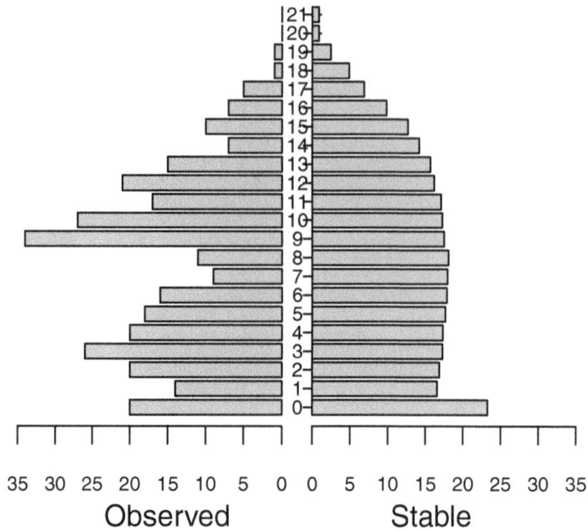

Fig. 9.2 Observed and stable age distribution in female snow leopards

distribution. The urn shape of the stable age distribution indicates a declining population (see Chap. 7).

Chapter 11 will discuss effects of age distribution on actual population growth versus estimated growth, and the gradual approach towards a stable age distribution.

9.5 Breeding Season

The equations as described above assume that all individuals at the start of each age class produce the average number of offspring (fecundity rate) for that class. This assumption holds for the birth–pulse breeding model in which the time of breeding and the time individuals enter a higher age class are the same.

Reproduction in a birth–flow model is spread over the year (age class). This means that some individuals will have died without producing offspring. The number of births in that age class will be overestimated, resulting in higher values for the measures mentioned, if this mortality is not taken into account.

A general approach to dealing with this is to consider the number of potential breeders after 50 % of the mortality for that age class has occurred. The midpoint survivorship (L_x, see Eq. 7.7) is used to estimate the proportional number of births per age class i.e. $L_x m_x$.

9.6 Life Tables and Census

Imaginary cohorts that combine cohorts (in time) involve overlapping generations. Survival and fecundity may differ between (genetic) generations, e.g. due to inbreeding. Moreover, management procedures may have changed during the study period, e.g. reducing reproduction when the studbook population is reaching the carrying capacity.

Survival and fecundity rates in imaginary cohorts are averages[1] of the combined cohorts. This means that life tables may not necessarily represent the most recent trends, as mentioned earlier. It is recommended to compare growth rates as estimated from life tables with census data in the studied period. Although age distribution is not considered in census, differences in (annual) growth and birth rates provide information about trends in (annual) cohorts.

Figure 9.3a presents annual census data of the female snow leopard population in captivity for the period 1983–2002. The female population increased during the period 1983–1995 and then fluctuates around 300 individuals. The arithmetic mean growth rate for the same period is 4.2 % ($\bar{\lambda} = 1.042$). This positive growth is to be expected as the female population increased from 133 to around 300 individuals within the studied period.

This positive $\bar{\lambda}$ from census data differs from the negative growth as estimated from life table data over the same period. This discrepancy can be explained by the use of an imaginary cohort that combines data of 20 real (annual) cohorts. The age structure of the population has changed over time from a pyramid shape in 1983

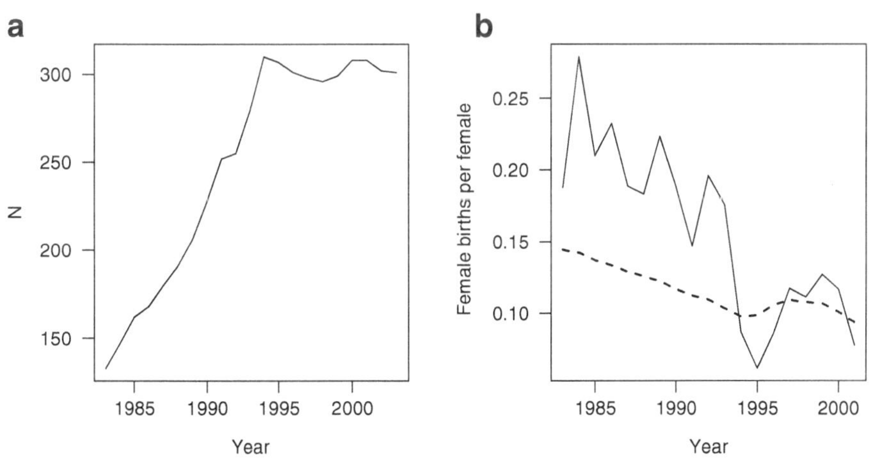

Fig. 9.3 (**a**) Annual census of female snow leopards during the period 1983–2002 (**b**) Female births per female. The *dotted line* refers to the geometric mean

[1]Weighted for individuals at risk in sub–cohorts.

to an ageing population in 2002 (see Fig. 7.2). Figure 9.3b shows that this change could be the result of managing population size by reducing reproduction i.e. the number of female births per female has declined. Since the population increased, since 1983, the contribution of recent cohorts to the imaginary cohort outweighs the smaller early cohorts. This means that mortality and fecundity rates, and the derived growth rates, are more likely to reflect the recent period.

Annual cohorts in studbook populations are generally small. For example, the annual number of female births (cohorts) in the snow leopard population ranges from 19 to 50 during the period 1983–2002. Each of these cohorts is too small for life table analysis. This means that creating an imaginary cohort for this studbook population is unavoidable. The study period as used in this example is selected to cover the lifespan of approximately 21 years in this species (see Table 7.3). This relatively long period embeds the risk that the imaginary cohort consists of cohorts with different mortality and/or fecundity rates. Therefore, it is recommended to include information on trends observed in the census data when selecting an appropriate study period.

Selection of a study period will undoubtedly be a compromise between observed trends and available data to construct a sufficiently large cohort. Moreover, the available studbook data may not cover the (potential) lifespan. For example, restricting a study period for female snow leopards to the period 1995–2002, when female population size is stable, only covers a third of the potential lifespan. Although left truncation results in including snow leopards born before 1995 (and thus older ages), it also introduces individuals that may have been exposed to different conditions.

References

Ballou JD, Lacy RC (1995) Identifying genetically important individuals for management of genetic variation in pedigreed populations. In: Ballou JD, Gilpin M, Foose TJ (eds) Population management for survival and recovery: analytical methods and strategies in small population conservation. Methods and cases in conservation science series. Columbia University Press, New York, pp 76–111

Ballou JD, Lees C, Faust LJ, Long S, Lynch C, Bingaman Lackey L, Foose TJ (2010) Demographic and genetic management of captive populations. In: Kleiman DG, Thompson KV, Kirk Baer C (eds) Wild mammals in captivity: principles and techniques for zoo management, 2nd edn. The University of Chicago Press, Chicago, pp 219–252

Ballou JD, Lacy RC, Pollak JP (2011) PMx: software for demographic and genetic analysis and management of pedigreed populations (version 1.0). Technical report, Chicago Zoological Society, Brookfield

Begon M, Harper JL, Townsend CR (2006) Ecology: from individuals to ecosystems, 4th edn. Blackwell Publishing, Malden

Fisher RA (1930) The genetical theory of natural selection, 3rd edn. Clarendon Press, Oxford

Galindo C (2008) On Fisher's reproductive value and Lotka's stable population. In: European population conference, Barcelona, 8–12 July 2008. European Association for Population Studies, Spain

Krebs CJ (1994) Ecology: the experimental analysis of distribution and abundance, 4th edn. Addison–Wesley, San Francisco
Lotka AJ (1922) The stability of the normal age distribution. Proc Natl Acad Sci 8(2):339–345
Mertz DB (1970) Notes on methods used in life–histories. In: Connell JH, Mertz DB, Murdoch WW (eds) Readings in ecology and ecological genetics. Harper and Row, New York, pp 4–17
Princée FPG (2014) Population Management Library (PML) software and manual version 1.0. East Rudham. http://www.princee.com

Chapter 10
Survival Analysis

Abstract The problem of censored data is not unique to studbooks, but was already recognised in medical trials, e.g. patients who withdraw from a study. The non–parametric Kaplan–Meier Product Limit Estimator which estimates survivorship (l_x) in flexible intervals, i.e. each time a death occurs or an individual leaves the population (alive), was developed. Since dates of death and lost to follow–up are recorded in studbooks, the Kaplan–Meier estimator adds a finer detail to survival analysis than age–class based methods. This method is used to study effects of treatments and statistical tests have been developed to compare results. This means that the Kaplan–Meier estimator can be used to study differences in subgroups, e.g. inbred/non–inbred groups. Statistical tests can be sensitive to the shape of survival curves i.e. early or late deaths. The Q-test to assist in selecting the appropriate test is described. The semi–parametric Cox Proportional Hazard Regression model is an extension of the Kaplan–Meier estimator that is particularly suitable for studying the effects of numerical measures, e.g. inbreeding coefficients. Effects can be predicted in terms of increase/decrease in survival per unit. The result of Cox's model is presented as a hazard ratio which can be transformed to the more intuitive survivorship. This model can handle pseudoreplication (frailty), left truncated and recurrent data. The Kaplan–Meier estimator is versatile and can be applied to any one–time event. Its use in age at first breeding is demonstrated with data from the snow leopard studbook.

10.1 Introduction

The problem of incomplete observations, either because individuals live beyond, or are lost to follow–up during, the study period, is certainly not unique to studbook data (see Sect. 7.5.3). The medical world is confronted with similar problems, for example, in clinical trials of different treatments (Kaplan and Meier 1958). Some patients may die during the study period, others live beyond the end, and some patients may not return for further check–ups i.e. they are lost to follow–up.

F.P.G. Princée, *Exploring Studbooks for Wildlife Management and Conservation*, Topics in Biodiversity and Conservation 17, DOI 10.1007/978-3-319-50032-4_10

The *Kaplan–Meier product limit estimator* (Kaplan and Meier 1958) (shortened to *Kaplan–Meier estimator*) is a non–parametric method that calculates survival in the cohort at the time (age) when a member dies or is censored. It thus differs from survival in fixed intervals (age classes) as discussed in Chap. 7.

The Kaplan–Meier estimator is a versatile method that was developed for any study in which time until a (random) event occurs in a group of objects (products) is measured, e.g. a manufacturer who studies the proportion of failing incubators during the guarantee period; or the age at first breeding (see Sect. 10.5). Any time unit that suits the study, from nanosecond to galactic year, can be used.

It may not be a surprise that the Kaplan–Meier estimator is also applied in ecology and population biology, such as for survival studies using radio–telemetry (Krebs 1989; Pollock et al. 1989b), or survival studies on zoo populations (Debyser 1995; Nuss and Warneke 2009). Recent studbook analysis software programs, e.g. PMx (Ballou et al. 2011), PopLink (Faust et al. 2012) and PML/studbookR (Princée 2014a,b) have implemented this method to create survivorship curves.

Survivorship curves in management of (zoo) populations tend to be used mainly in population projections (see Chap. 11). However, several nonparametric statistical tests, such as the Mantel–Haenszel or logrank test in Chap. 7, have been developed to compare survival distributions (see Harrington and Fleming 1982; Martinez and Naranjo 2010). The combination of these tests and survivorship curves based on (daily) events opens a "wealth" of detailed comparative studies, e.g. effects of husbandry practices on neonatal survival, differences between managed and non–managed populations, or effect of age on survival time since transfer to a new location.

The semi–parametric *Cox proportional hazards regression model* extends the Kaplan–Meier estimator with regression analysis (Cox 1972). Whereas the Kaplan–Meier estimator is restricted to hypothesis testing between categorical groups (e.g. sex or type of rearing), Cox's model can handle numerical variables (e.g. inbreeding coefficients, age of dam) and, moreover, quantifies effects by estimating proportional differences in hazard rates (see also discussion of hazard rates in Sect. 8.6).

Survival data can be subject to pseudoreplication (Hurlbert 1984), e.g. the litter effect in species with multiple births and institutional differences in husbandry practices. Software implementations of the Cox's regression model can handle pseudoreplication (e.g. Therneau 2015a,b).

The Kaplan–Meier estimator and the Cox's regression model are both implemented as standard in various statistical software e.g. in the R package *survival* (Therneau 2015a) which is used in this book.

10.2 Kaplan–Meier Product Limit Estimator

The Kaplan–Meier Product Limit Estimator is used as a standard tool estimating survival at time t (symbol: S_t) in clinical trials (Klein and Moeschberger 1997):

$$S_t = \prod_{t_i \leq t} \left[\frac{N_i - D_i}{N_i} \right] \quad (10.1)$$

where N_i and D_i is the number at risk (of death) and number of (actual) deaths, respectively, at time i (Kaplan and Meier 1958).

Survival in Kaplan–Meier estimator is similar to age–class based survivorship in life tables (see Eq. 7.6), i.e. when the "study" starts at birth. The software programs PMx (Ballou et al. 2011) and PopLink (Faust et al. 2012) can transform Kaplan–Meier survival times to age–class based survivorship.

Figure 10.1 presents survivorship and its 95 % confidence intervals for female Californian sea lions in the European studbook (data 1978–2007). Survivorship at birth is per definition 1.0, as in age–class analysis. However, stillbirths and deaths at date of birth are considered "events", resulting in an l_0 of less than 1.0.

Summary statistics for survivorship in female Californian sea lions are presented in Table 10.1. The median life expectancy is 4,746 days ($\approx$13 years) with a wide 95 % confidence interval from $\approx$9.5 to 19.2 years. The mean life expectancy (3,461 days) and its standard error are estimates, as the oldest individual was still living at the end of the study (see Therneau 2015a).

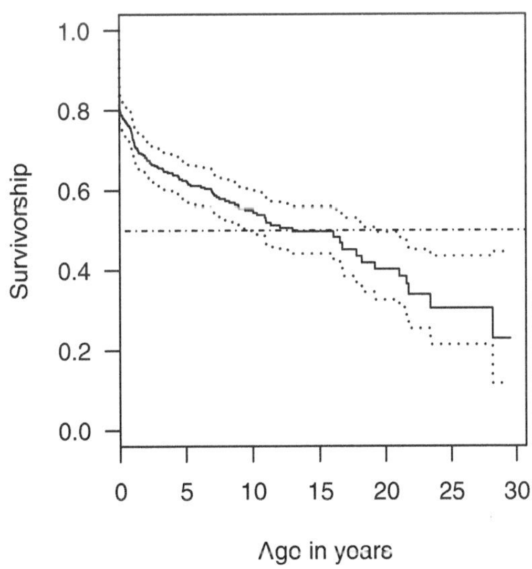

Fig. 10.1 Survivorship (l_t) curve (*solid line*) in female Californian sea lions born in the European studbook population between 1978 and 2007 ($N_f = 453$, $D = 197$). *Right* censoring has been applied. The *dotted lines* represent the 95 % confidence intervals. The X axis is re–scaled to years

Table 10.1 Summary of Kaplan-Meier statistics on female survivorship in Californian sea lions that were born in the European studbook population between 1978 and 2007 ($N_f = 453$, $D_f = 197$). The data are right–censored

	95 % Confidence interval of median			
Median	Lower	Upper	Mean[a]	SE_{mean}
4,746	3,461	6,996	5,172	289

N_f number of females, D_f female deaths, *SE* standard error of the mean
[a] Restricted mean with upper limit of 10,784 days (29.5 years)

Right–censored individuals are included in being at risk, either until the end of the study period or until the day they are lost to follow–up. This means that prorating (see Chap. 7) is not an issue in analysis based on the Kaplan–Meier estimator.

10.2.1 Statistical Tests

Kaplan–Meier survival curves for male and female blesbok born in the European studbook population between 1970 and 2008 are presented in Fig. 10.2. The median lifespan of females is 6 years and 9 months (2,467 days). This is more than twice the median lifespan of 3 years and 2 months (1,154 days) in male blesbok. Given the relatively large sample sizes, i.e. $N_f = 430$ and $N_m = 451$, one may expect that this difference is significant.

The Mantel–Haenszel or logrank test, as described in Chap. 7, is also applied in survival analysis (Klein and Moeschberger 1997). The result of this test applied to the blesbok example is $\chi^2 = 24.6$, $p < 0.0001$, which means that the null hypothesis that survival in males and females is the same is rejected.

Several variations (and names) of rank tests have been developed (see Léton and Zuluaga 2005). The *Gehan–Wilcoxon test* is also frequently used in survival analysis (Martínez et al. 2005). The logrank test is an unweighted test and is more sensitive for differences in distributions at older ages, while the Gehan–Wilcoxon test is a weighted test that places more emphasis on differences in younger ages (see Klein and Moeschberger 1997; Martinez and Naranjo 2010).

Figure 10.2 shows that differences between male and female survivorship are larger at older ages (but not the oldest ages). This suggests that the logrank tests would be appropriate. The *Q test* is a pretest that has been designed to assist in the selection of either the logrank or Gehan–Wilcoxon test (Martinez and Naranjo 2010). The procedure is as follows:

1. Identify the lower survival curve (S_m); which is the male curve in the blesbok example.
2. Determine the ages (time t) nearest to $S_m(t) = 0.2$ and $S_m(t) = 0.6$ from the lower survival curve. For the (male) blesbok $t(0.6) = 704$ and $t(0.2) = 3,413$ (in days).

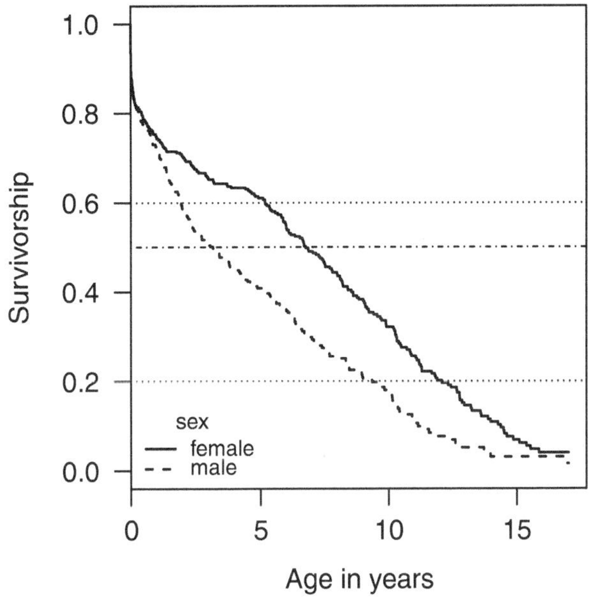

Fig. 10.2 Right censored survivorship curves for female (*solid line*) and male (*dashed line*) blesbok, born in the European studbook population between 1970 and 2008 ($N_f = 430$, $D_f = 284$; $N_m = 451$, $D_m = 300$). The *dotted-dashed horizontal line* is the median survivorship. The *dotted horizontal lines* refer to survivorship of 0.2 and 0.6. The X axis is re–scaled to years

3. Determine the survival rates in the upper survival curves that correspond to nearest ages as found for males in (2); for the female blesbok these are $S_f(704) = 0.71$ and $S_f(3,413) = 0.35$.
4. Estimate:
   $Q = (\, S_f(704) - S_m(704)\,) - (\, S_f(3,413) - S_m(3,413)\,)$
   $Q = (0.71 - 0.6) - (0.35 - 0.2)$
   $Q = -0.04.$
5. The logrank test is used when $Q < 0$, otherwise the Gehan–Wilcoxon test is used.

Since $Q < 0$ the logrank test is recommended for testing the hypothesis that survival is the same in males and females. The dotted horizontal lines in Fig. 10.2, which represent both survival rates 0.6 and 0.2, can be used to approximate Q.

10.3 Cox Proportional Hazards Regression

The rank tests discussed above evaluate differences in distribution patterns (shape of curve) of survival between groups. This means that one cannot formally conclude that values for the median lifespan differ. For example, survival between sexes may start to differ after the age of sexual maturity. In species with large litter sizes and high early mortality, 50 % in both sexes may have died before reaching sexual maturity.

Statistical tests for comparing medians of survival curves have been developed, but they are sensitive to type I errors (Chen and Zhang 2016). Therefore, these tests are not discussed in this book.

Quantifying differences between survival curves can be achieved with parametric models as discussed in Chap. 7 or the semi–parametric *Cox proportional hazards regression model* (Cox 1972) (shortened to *Cox's model*).

Cox's model is an extension of the Kaplan–Meier estimator and similar to a multiple regression where the response is the hazard (symbol: h) (see Sect. 8.6). This regression is written in the form:

$$log_e\big(h(t)\big) = log_e\big(h0(t)\big) + \beta_1 x_1 + \beta_2 x_2 \ldots + \beta_p x_p \tag{10.2}$$

where $h(t)$ is the hazard at time t; $x_1, x_2 \ldots x_p$ are the explanatory variables, e.g. sex, inbreeding coefficient, age of dam; and $h0(t)$ is the baseline hazard when all the explanatory variables are zero; $\beta_1, \beta_2 \ldots \beta_p$ are regression coefficients.

The Cox's model does not assume an underlying distribution, such as the Gompertz–Makeham or Weibull distribution, but estimates the *hazard ratio (HR)* between categories (e.g. males and females) or per unit (e.g. age of dam, litter size) without estimating the base line hazard ($h0(t)$).

The mathematics behind the Cox's model are more complicated than those of the Kaplan–Meier estimator. Therefore, we leave this aspect with the quote "The coefficients are estimated from the data using a statistical package" (Bewick et al. 2004), and move on to the interpretation of results. See Klein and Moeschberger (1997), Kleinbaum and Klein (2006) and Klein et al. (2013) for more in–depth details on Cox's model. This book uses the software implementation of Cox's model in the R package survival.

Table 10.2 presents results of Cox's model on survivorship differences between female and male blesbok. In R/survival the first group in alphabetical order or factor levels within categorical data is used as the base–line.[1] This means that only results for the male group in relation to the base–line (female) group are provided.

Table 10.2 Summary of Cox proportional hazards regression model on differences in survivorship between sexes in blesbok, born in the European studbook population between 1970 and 2008 ($N_f = 430$, $D_f = 284$, $N_m = 451$, $D_m = 300$). The data are right–censored. Female survival is used as base–line (see text)

Predictor	β	SE_β	z	$p > \|z\|$	HR	95 % Confidence	Model p
Male	0.325	0.075	4.347	0.000	1.385	[1.196, 1.603]	0.527

N_m males, D_m male deaths, N_f females, D_f female deaths, β regression coefficient, SE_β standard error of β, z Wald statistic, $p > |z|$ probability of z, *HR* hazard ratio, *Model p* Therneau–Grambsch test

[1]The order can be changed by manually assigning levels to factors in R.

The main value of interest is the hazard ratio (*HR*), which is estimated from the regression coefficient (β) as:

$$HR = e^{\beta} \tag{10.3}$$

The *HR* value of 1.385 in Table 10.2 means that males have a 1.385 times (38.5%) higher risk of dying (hazard) at any time (age) than females of the same age. An important assumption of the Cox's model is that the hazard ratio between categories is constant over time (age), hence the previous phrase "hazard at any time".

The easiest way to check whether this assumption of constant hazard ratio is met, is to compare survival curves of categories (in this case sex). Whenever two or more categories cross, the assumption is not met. The survival curves of male and female blesbok do not cross (Fig. 10.2). This means that a statistical test such as a goodness–of–fit test (Kleinbaum and Klein 2006) or the *Therneau–Grambsch test* (Grambsch and Therneau 1994) is required to test the assumption of constant hazard ratio. The second test is implemented in R/survival (Therneau 2015a).

The null hypothesis is that the hazard ratio is constant over time (the *proportional hazards assumption* or PH assumption). The test result can be somewhat counter–intuitive as the (Cox) PH assumption is *not* met when $p < 0.05$. The hazard ratio between male and female blesbok is considered constant i.e. $p = 0.527$ (Table 10.2).

A hazard ratio that is constant over time does not indicate whether the differences between categories (e.g. male and female) – expressed as regression coefficient(s) – is/are significant. The *Wald test* can be applied to test whether the regression coefficient β differs significantly from 0 (the null hypothesis). The result of this test is $z = 4.347$, $p < 0.0001$, and therefore the null hypothesis is rejected.

A significant regression coefficient alone is not sufficient to accept the Cox's regression model. The confidence interval for the hazard ratio should not include 1 (as this would mean that male and female survivorship are not different).

The Kaplan–Meier estimator is not really suitable for analysing effects of numerical variables, such as age of the parent(s), year of birth (cohort) and inbreeding coefficients, on survival. For example, each observed cohort would be considered as a separate category. The Cox model can handle numerical variables. This section will use cohorts of Nepalese red pandas as an example.

Survival analyses in this chapter (and Chap. 7) are based on imaginary cohorts (Caughley 1977). The choice of combining cohorts is a balance between sufficient sample size and similar husbandry conditions (see discussion in Chap. 3). Cox's model can be applied to test for differences between cohorts e.g. to test whether there is a trend of increased survival over time.

Table 10.3 presents results on survival differences in real cohorts of female Nepalese red pandas from 1977 to 2012. The different "cluster" models are ignored for the moment. Cox's model estimates the hazard ratio of continuous variables per unit, which is 1 year for red panda cohorts. The estimated hazard ratio of 0.98, which is significant, means that the risk of females to die at any time, has declined with $\approx$2 % per year since 1977.

Table 10.3 Summary of Cox proportional hazard regression on cohort differences in survivorship under different marginal (cluster) models in female Nepalese red pandas born between 1977 and 2012 ($N = 926$, $D = 665$). The data are right–censored

Cluster	β	*Robust* SE_β	z	$p > \|z\|$	HR	95% Confidence	Model p
	−0.019	0.0048[a]	−4.035	0.0001	0.981	[0.9719, 0.9902]	0.842
Litter	−0.019	0.0049	−3.856	0.0001	0.981	[0.9718, 0.9907]	0.840
Dam	−0.019	0.0051	−3.748	0.0002	0.981	[0.9716, 0.9910]	0.835
Site[b]	−0.019	0.0063	−3.037	0.0024	0.981	[0.9689, 0.9932]	0.791

N individuals, *D* deaths, β regression coefficient, *Robust* SE_β robust standard error, z Wald statistic, $p > |z|$ probability of z, *HR* hazard ratio, *Model p* Therneau–Grambsch test
[a]Standard error from default method
[b]Location of birth

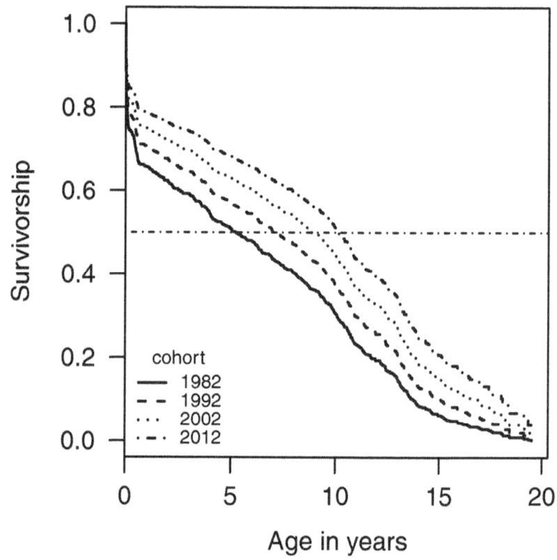

Fig. 10.3 Adjusted survivorship curves for female Nepalese red pandas born in the years 1982, 1992, 2002 and 2012. The X axis is re–scaled to years

Hazards are not straightforward to interpret. The trend in survival can be better observed by converting hazards into survival rates (see next section for method) to plot *adjusted survival curves* (i.e. adjusted for variables (Kleinbaum and Klein 2006)). Figure 10.3 presents adjusted survival curves for the female cohorts 1982, 1992, 2002 and 2012 as predicted from Cox's regression model. It shows the impact of a 2 % decrease in hazards per year on survival. For example, female median lifespan increased from ≈5 years (1,901 days) in 1982 to 10 years (3,690 days) in 2012.

It is important to realise that adjusted survival curves are predictions. Extrapolations to values for variables that lie outside the original data are not recommended. Such predictions, depending on the type of variable, may not even make sense.

Adjusted survival curves of categorical variables can be compared with those estimated with the Kaplan–Meier estimator. This provides an idea of the extent to which the predictions are reasonable. This will not work for numerical variables, as discussed above.

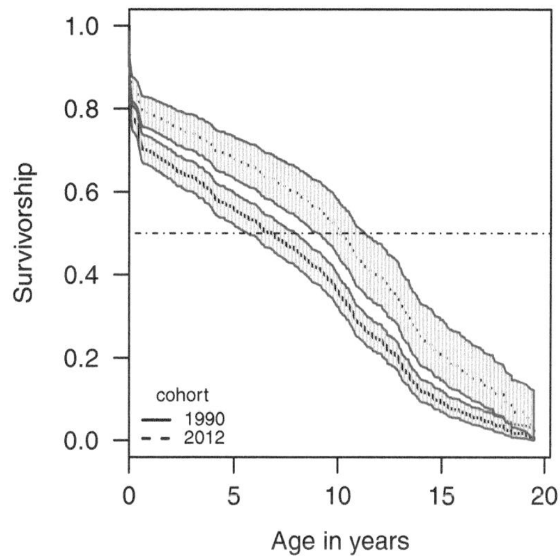

Fig. 10.4 Confidence limits of adjusted survivorship curves of female Nepalese red pandas born in the years 1982 (*solid line*) and 2012 (*dashed line*). The *horizontal* (*dash–dot*) *line* refers to 50 % survivorship. *Shaded bands* are 95 % confidence intervals. The X axis is re–scaled to years

Confidence intervals of survivorship can assist in assessing predictions. Figure 10.4 shows the 95 % confidence bands for adjusted survival curves of red pandas born in 1990 and 2012. Note that the confidence band for the 2012 data is considerably wider than for the 1990 cohort. This means that the predictive value for the 2012 cohort is low. Given that only 1 year of data on the 2012 cohort is available, such a low prediction should not be a surprise.

10.3.1 Frailty

The observed variance of survival data can be greater (*overdispersed*) than expected from a binomial distribution (see Eq. 8.1). This means that there is an additional random variation on top of the basic random (binomial) variation. In survival analysis, this additional random effect is named *frailty* (Vaupel et al. 1979), referring to the effect that the most "frail" individuals, i.e. those that are affected more by this random factor, may die early. Frailty models have been developed to handle overdispersion in survival analysis (Hougaard 1995; Wienke 2010).

A special type of frailty model is the *shared frailty model* of the proportional hazards regression (Cox's) model (Klein and Moeschberger 1997). This model refers to random effects that are shared by individuals in subgroups or clusters e.g. litter, family, zoological garden or protected area. In other words, shared frailty models can handle pseudoreplication (Hurlbert 1984) in survival data.

Shared frailty models estimate additional variance within identified subgroups, e.g. litters, through random effects that follow an assumed statistical (error) distribution such as gamma, Gaussian (normal) or negative binomial (Hougaard 2013). Another method is to use *generalized estimating equations (GEE)*, to estimate the additional variance due to differences between subgroups as observed in the data. The first method is a *frailty model*, the second method is a so-called *marginal model*. Both models are implemented in the R package *survival*.

Table 10.3 presents results of applying marginal models ("cluster") for litters, mothers (dam) and location of birth in survival analyses as described in the previous section. The regression coefficient (β) and consequently the hazard ratio do not differ between "plain" and marginal models. The robust standard error, however, has increased in marginal models compared to the plain model. Location of birth has more impact on the total variance than maternal and litter effects, which results in a wider confidence interval.

The results of analyses with marginal models do not change the conclusion that the year of birth (cohort) has a positive effect on female survival in Nepalese red pandas in captivity, i.e. p values are less than 0.005. However, this example shows that pseudoreplication can increase variance, and therefore increases the risk that the null hypothesis (no differences) is falsely rejected in the plain model.

The Cox's regression model in the survival package facilitates the use of a single variable in either the frailty model or the marginal model. However, there are cases where one would like to include multiple random effects (which are not necessarily associated with subgroups). Multiple random variables with Cox's model can be analysed, for example using the R package *coxme* (Therneau 2015b).

Note that categorical variables with large numbers of categories are unworkable models, as a regression coefficient for each category needs to be estimated. An example of an unworkable category would be the use of zoological institutions in an international studbook.

10.3.2 From Hazard Ratio to Survivorship

Although most statistical software will create survival curves from hazard ratios, it is important to understand some of the basic methods.

Hazard ratios for categorical variables are fixed values for each category compared with the baseline hazard; see the hazard ratio for males compared to females in Table 10.2 as an example. The hazard ratio for a continuous variable, like cohorts, is generally presented for one (1) unit. In the case of cohorts this unit is 1 year.

It can be confusing when the hazard ratio refers to one unit from a variable that ranges from 0 to 1. For example, in the case of inbreeding coefficients, the hazard ratio compares hazards between non inbred ($F = 0$) and fully inbred ($F = 1$) individuals. This can lead to very large values, e.g. $HR = 67.3$ in inbreeding effects on 6 month survival of Nepalese red pandas (see Chap. 13). In such cases it makes more sense to recalculate HR for intervals that are within the observed range, e.g. 0 to maximum observed inbreeding coefficient.

Equation 10.3 can be adjusted to facilitate the use of intervals other than 1 (including fractions):

$$HR_i = e^{\beta \times i} \tag{10.4}$$

where β is the regression coefficient in the Cox model, and i is the interval from 0 (the baseline hazard).

The regression coefficient in the above example is $log_e(67.3) = 4.21$. The hazard ratio for $F = 0.25$ would be $e^{4.21*0.25} = e^{1.05} = 2.86$. This means that the hazard of dying at any time is 2.86 times higher in the inbred $F = 0.25$ group than in the non–inbred group.

Knowing the hazard ratio allows the estimation of survival for interval i at any time (age) using the following equation:

$$S_i(t) = S_0(t)^{HR_i} \tag{10.5}$$

where $S_0(t)$ is the baseline survival and $S_i(t)$ survival at interval i.

However, there is a caveat in this method as the Cox's model calculates the hazard ratio without "knowing" the exact distribution of the baseline hazards $h_0(t)$. Estimation of the empirical cumulative hazard $H_0(t)$ is one of the methods to estimate $S_0(t)$ (Klein and Moeschberger 1997), which is based on the following relation:

$$S(t) = e^{-H(t)} \tag{10.6}$$

Statistical software, such as R/survival (Therneau 2015a), is needed to estimate the cumulative hazard, especially from larger sample sizes.

10.4 Left Truncated and Recurrent Data

The survival data in the previous section were right censored, but not left truncated. The main reason is that versions of logrank tests that are implemented in basic packages in statistical software, e.g. survival (Therneau 2015b) and SAS (Foreman et al. 2008) are not valid under left truncation. This practical limitation also refers to *recurrent data*, i.e. individuals that are not at risk during particular periods of time e.g. during the time an individual is on loan outside the studied zoo region. Such an individual will be considered at risk as soon as it returns. This means that differences between Kaplan–Meier survival curves can not be tested with the standard packages.

Left truncated, right censored and/or recurrent survival data can be analysed with the Cox's model. This does not mean that the Kaplan–Meier estimator can not be used for testing differences in these types of survival data. Logrank tests that are adjusted for left truncated data have been developed (Pollock et al. 1989a). Packages that can handle recurrent data also been developed, e.g. R package *survrec* (González et al. 2013).

10.5 Age at First Breeding

Although this chapter is dedicated to survival analysis, some other uses of the above methods in the field of demography will be briefly introduced in this section. The Kaplan–Meier estimator is, as mentioned in the introduction, not restricted to estimating survivorship. In principle, any one–time event (per individual), such as time until *first* breeding, can be studied. The original name product–limit estimator is in that sense more appropriate. The measure in this analysis refers to the proportion of individuals that have *not* reproduced (i.e. the "non–breeders"). Right censoring in analysis of first age at reproduction applies to non–breeders in the cohort which have died (they cannot reproduce), or are lost to follow–up during the study period, or are living at the end of the study period.

Figure 10.5 presents the proportion of non–breeders in female (solid line) snow leopards that are born in the period 1983–2003. Summary statistics for both sexes are presented in Table 10.4. The proportion of non–breeders stays 1 until the first birth has occurred around the age of 2 years (note that the analysis in Chap. 3 resulted in a younger age, i.e. 500 days, due to an outlier/artefact in the data).

Fifty percent of the females and males have reproduced at least once at the (median) age of ≈7 and 8 years (2,571 and 2,869 days), respectively (Table 10.4). The mean age at first reproduction is an estimate depending on whether individuals living at the end of the study period reproduced or not.

The results of the logrank test imply that the age at first breeding curves of males and females do not differ ($\chi^2 = 0, p = 0.92$). This also means that the median age at first breeding does not differ between sexes in snow leopards.

The proportions of non–breeders do not decline at the tail of the distribution as individuals have reached the post–reproductive lifespan. Some 38 % and 30 % of

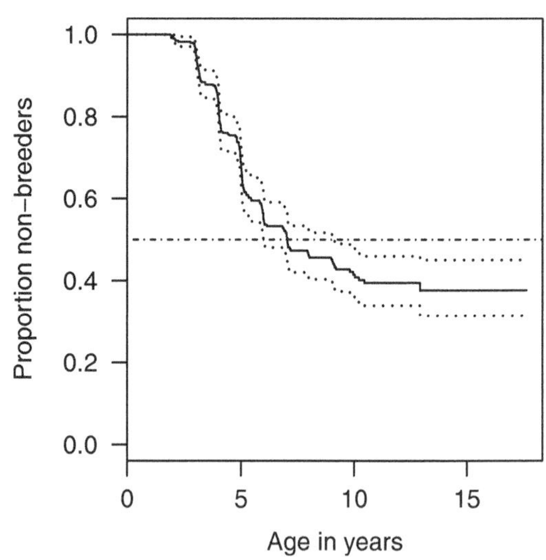

Fig. 10.5 Proportion of non–breeders in female snow leopards that are born between 1983 and 2003; right censoring is applied. The *dotted lines* are the 95 % confidence intervals. The *horizontal line* represents the median first age at birth. The X axis is re–scaled to years

Table 10.4 Summary of product–limit statistics for age at first reproduction (in days) in female and male snow leopards that are born between 1983 and 2003

			95 % confidence interval of median			
Sex	n	Median	Lower	Upper	Mean[a]	SE_{mean}
Female	697	2,571	2,186	3,319	3,542	130
Male	674	2,869	2,544	3,306	3,451	133

n number of individuals, SE_{mean} standard error of the mean
[a]Restricted means with upper limits of 6,104 and 5,785 days in females and males, respectively

female and male snow leopards in the cohort, respectively, have never reproduced. This important information is not available from fecundity tables (Chap. 7).

However, one needs to be careful in interpreting the proportion of non–breeders in an imaginary cohort. Individuals that have not reached sexual maturity by the end of the study period, e.g. snow leopards born in 2002 and 2003 in the example, lead to overestimation of the proportion of non–breeders. It is recommended to carry out the analysis of first age at breeding twice, i.e. with younger cohorts to estimate the youngest (social) breeding age, and without to estimate the proportion of non–breeders.

Analysis of age at first breeding introduces a different type of censoring i.e. *left censoring*. Left censoring refers to individuals to which the event, e.g. birth, has occurred before the study period, whereas in left truncation the event has *not* happened (which explains why left censoring, logically seen, can not be applied in survival analysis).

The above means that one needs to ensure that left truncation in analysis of age at first birth is not left censoring. This would either wrongly increase the proportion of non–breeders or increase age at first birth by considering the second or later births within the study period as first births.

References

Ballou JD, Lacy RC, Pollak JP (2011) PMx: software for demographic and genetic analysis and management of pedigreed populations (version 1.0). Technical report, Chicago Zoological Society, Brookfield

Bewick V, Cheek L, Ball J (2004) Statistics review 12: survival analysis. Crit Care 8(5):389–394

Caughley G (1977) Analyses of vertebrate populations. The Blackburn Press, Caldwell

Chen Z, Zhang G (2016) Comparing survival curves based on medians. BMC Med Res Methodol 16(1):1–7. doi:10.1186/s12874-016-0133-3

Cox DR (1972) Regression models and life–tables. J R Stat Soci Ser B (Methodol) 34(2):187–220

Debyser IWJ (1995) Juvenile mortality in captive populations of primates: an epidemiological and pathological study in Dutch and Belgian zoological gardens and a German primate centre. PhD thesis, University of Utrecht, Faculty of Veterinary Medicine, Utrecht

Faust LJ, Bergstrom YM, Thompson SD, Bier L (2012) PopLink version 2.4. Lincoln Park Zoo, Chicago

Foreman AJ, Lai GP, Miller DP (2008) Surviving left truncation using PROC PHREG. In: 2008 annual conference proceedings, Western Users of SAS Software, Universal City
González JR, Peña EA, Strawderman RL (2013) survrec: survival analysis for recurrent event data. http://CRAN.R-project.org/package=survrec. R package version 1.2-2
Grambsch PM, Therneau TM (1994) Proportional hazards tests and diagnostics based on weighted residuals. Biometrika 81(3):515–526
Harrington DP, Fleming TR (1982) A class of rank test procedures for censored survival data. Biometrika 69(3):553–566
Hougaard P (1995) Frailty models for survival data. Lifetime Data Anal 1(3):255–273
Hougaard P (2013) Frailty models. In: Klein et al. (2013), pp 458–473
Hurlbert SH (1984) Pseudoreplication and the design of ecological field experiments. Ecol Monogr 54(2):187–211
Kaplan EL, Meier P (1958) Nonparametric estimation from incomplete observations. J Am Stat Assoc 53:457–481
Klein JP, Moeschberger ML (1997) Survival analysis: techniques for censored and truncated data, 1st edn. Statistics for biology and health, Springer, New York
Klein JP, van Houwelingen HC, Ibrahim JG, Scheike TH (eds) (2013) Handbook of survival analysis. Chapman and Hall/CRC, Boca Raton
Kleinbaum DG, Klein M (2006) Survival analysis: a self–learning text. Statistics for biology and health, 2nd edn. Springer, New York
Krebs CJ (1989) Ecological methodology. Harper Collins Publishers, New York
Léton E, Zuluaga P (2005) Relationships among tests for censored data. Biom J 47(3):377–387
Martinez RLMC, Naranjo JD (2010) A pretest for choosing between logrank and Wilcoxon tests in the two–sample problem. Int J Stat LXVIII(2):111–125
Martínez M, Rodríguez–Vigal C, Jones OR, Coulson T, Miguel AS (2005) Different hunting strategies for different weights in red deer. Biol Lett 1:353–356
Nuss K, Warneke M (2009) Life span, reproductive output, and reproductive opportunity in captive Goeldi's monkeys (*Callimico goeldii*). Zoo Biol 28:1–15
Princée FPG (2014a) Population Management Library (PML) version 1.0, software and manual. East Rudham. http://www.princee.com
Princée FPG (2014b) studbookR: Statistical analyses of studbooks version 0.2, software and manual. East Rudham. http://www.princee.com
Pollock KH, Winterstein SR, Bunck CM, Curtis PD (1989a) Survival analyses in telemetry studies: the staggered entry design. J Wildl Manag 53:7–15
Pollock KH, Winterstein SR, Conroy MJ (1989b) Estimation and analysis of survival distributions for radio–tagged animals. Biometrics 45(1):99–109
Therneau T (2015a) A package for survival analysis in S. http://CRAN.R-project.org/package=survival. R package version 2.38
Therneau TM (2015b) coxme: mixed effects Cox models. http://CRAN.R-project.org/package=coxme. R package version 2.2-5
Vaupel JW, Manton KG, Stallard E (1979) The impact of heterogeneity in individual frailty on the dynamics of mortality. Demography 16(3):439–454
Wienke A (2010) Frailty models in survival analysis. Biostatistics series, Chapman and Hall/CRC, Boca Raton

Chapter 11
Population Projections

Abstract Population projections based on observed growth trends are important in captive animal management. Several methods, which require different levels of detail in the data, are available. A projection based on the geometric mean of population growth (λ) estimated from recent census data, optionally per sex, is the simplest model. Ecological growth models, e.g. logistic growth that fits the population data, can be used to project future development of the population. Census data from the blesbok studbook are used to illustrate these census–based methods. The Leslie matrix is applied to project population development of age– and stage–structured populations. Application of this method is described for birth–pulse and birth–flow reproduction. Damping of λ and convergence to a stable age distribution are also described. Since studbook software handles projections, managers are generally not exposed to matrix algebra, so a brief introduction to matrix algebra is presented as this method is often applied in ecological studies. Matrix calculus provides methods for testing the effects of changing individual survival and fecundity rates on λ. Changes can be estimated as absolute values (sensitivity) or proportional values (elasticity). The standard Leslie matrix is a deterministic model. A stochastic version of this matrix model is used to show the effects of demographic uncertainty on population development. The differences between projection and prediction are discussed. A retrospective study in which projections are compared with actual population development is presented. The advantages and disadvantages of applying stage–based models to studbook populations are discussed.

11.1 Introduction

Chapters 4, 5, 6, 7, 8, 9 and 10 focus on methods and techniques to assess historical and current trends in studbook(–like) populations. Knowledge and understanding of population dynamics are essential in order to adapt and improve management for maintaining demographically viable populations. Being able to make *population projections* is an intrinsic part of this management.

The *Leslie matrix* (Leslie 1945, 1948) can be considered as the "standard" method for population projections using studbook software (see Chap. 2). The Leslie matrix not only projects population size, but projects age distribution, separately for males and females, as well.

F.P.G. Princée, *Exploring Studbooks for Wildlife Management and Conservation*, Topics in Biodiversity and Conservation 17, DOI 10.1007/978-3-319-50032-4_11

This method requires data on age–specific fecundity and mortality, and on age distribution. These data are not necessarily available for all studbook species. For example, age–specific fecundity rates can not be estimated when parentages are incomplete, such as in species that live in herds or colonies. The alternative method is to project population size based on census data (see Chaps. 4 and 6).

Small sample sizes result in high variances in age–specific fecundity and mortality rates (Chap. 7), which consequently can affect the quality of population projections. Comparison with projections based on other methods, such as annual growth rates in census data, is recommended to detect flaws in life tables.

11.2 Census Data

Chapters 4 and 6 described methods to estimate population growth rates from census data. These growth rates can be used to project population size.

The finite rate of increase (λ), in practice often expressed as the annual growth rate, provides a straightforward measure to project population size:

$$N_{t+1} = \lambda N_t \tag{11.1}$$

where N_t and N_{t+1} are the population sizes at times t and $t + 1$, respectively.

Since λ can fluctuate greatly per year, using the most recent value is not the best option for population projections. The geometric mean (λ_G) is a better measure to reflect a trend in population growth (see method in Chap. 4).

Figure 11.1a presents census data for the period 1997–2007, and a 25 year projection of the European studbook of blesbok. The geometric means for male, female and total population for this census period, i.e. 1.023, 0.998 and 1.009 respectively, were used.

The projected population develops as would be expected from the $\lambda_{G[1997,2007]}$ values. However, this does not mean that the projections are correct. The projected male population increases and surpasses the female population. That is very unlikely, given that the female population is declining, blesbok produce singletons, and the overall sex–ratio at birth in the studbook population is equal (binomial test, $p = 0.278$). This means that the male population will decline sooner rather than later. The population projection in Fig. 11.1a does not reflect the reality.

The projection of total population size based on $\lambda_{G[1997,2007]}$ reflects the trend as observed in male blesbok during the last 8–9 years, but less well in females, where the declining trend in the female population starts around 2001 (see Fig. 11.1a). It would therefore be more realistic to project the female population size based on this decline. Figure 11.1b presents the projected female population size based on λ_G for females in the period 2001–2007.

The female population is declining much faster than projected with $\lambda_{G[2001,2007]}$ (for the above reasons). These projections can be extended by extrapolation from the 95 % *confidence interval* for the geometric mean (Norris 1940). The projections

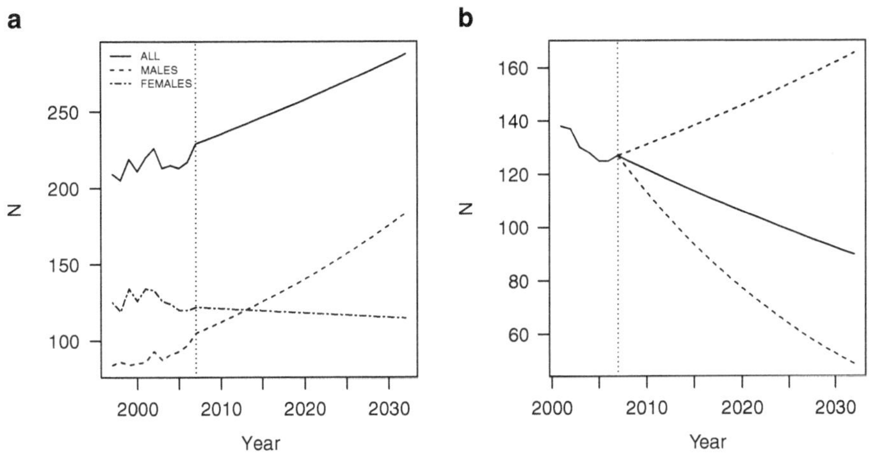

Fig. 11.1 (**a**) Census of blesbok population for the period 1997–2007 and 25 years projected size based on the geometric mean of λ for this census period. *dot–hashed line*: females; *hashed line*: males; *solid line*: males, females and unknown sex. (**b**) Female census for the period 2001–2007 and 25 years projected size based on the geometric mean of λ for this census period. The *dotted horizontal lines* refer to projections based on the 95 % confidence interval of the geometric mean

based on the upper and lower limits of the confidence interval for $\lambda_{G[2001,2007]}$ (0.96 and 1.01) show that the female population can also decrease more rapidly or even increase (see Fig. 11.1b). Given these differences one may wonder whether this projection has any value other than indicating the limitations of projections.

Fecundity and mortality rates are assumed to remain the same in the previous example. This means that the annual growth rate in the above model is assumed to remain constant during the projected period of time. Such an assumption is not likely to hold given the limited space that is available per species for captive breeding (Princée 1998).

Chapter 6 showed that a logistic growth model fits development of the female blesbok population during the period 1970–2007 ($\chi^2 = 23.97$, $p = 0.95$, $df = 37$). Projections can be made on the basis of this growth model. Figure 11.2 presents a 25 years projection of the female blesbok population.

This population grows until it has reached a size of 145 (as estimated by the model). This size would be the carrying capacity of the habitat for a free–ranging population which is subject to density–dependent growth. It may well be that the European studbook population of blesbok is reaching its carrying capacity, i.e. available zoo space, and that management measures are taken to control population growth. However, it would be naive to make such a conclusion based purely on the data without consulting studbook keepers. In the end, even a well–fitted logistic growth model is still a model.

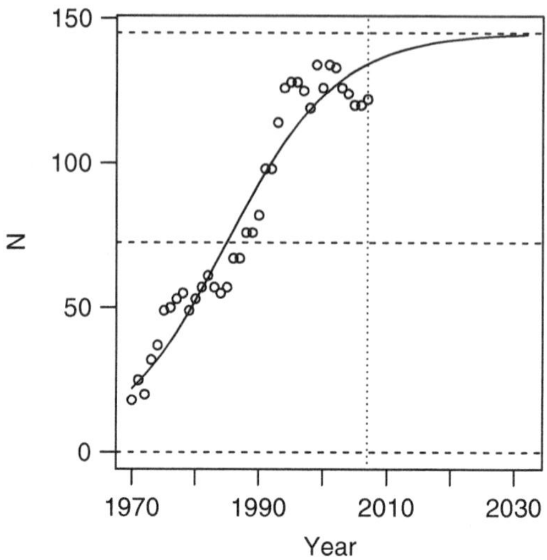

Fig. 11.2 Logistic growth model for the female blesbok population in the period 1970–2007 and 25 year projection (*solid line*). *Open circles* are census data for the period 1970–2007

11.3 Leslie Matrix

Population projections based on census counts and (annual) growth rates do not reveal the effects of unstable age distributions on population size. For example, the age distribution of the female snow leopard population in 2002 has more individuals in the older reproductive classes than in younger age classes (see Chap. 9). This implies that population size can decline, at least initially, instead of remaining more or less stationary in size, as indicated by the net reproductive rate of 1.009.

It is important to use age distribution and life table data, whenever available, in population projections and to study the effects of age distribution on population growth. The Leslie matrix (Leslie 1945, 1948) was developed to project age distribution and population size using these data. The projection (time) interval for age–structured populations is equal to the class–width.

The underlying matrix algebra of the Leslie model can be somewhat "intimidating", but the basic principles are straightforward: (1) individuals pass from one class to the next with the survival probability P_x; and (2) each individual in class x that survives until the start of the breeding season, is expected to produce m_x offspring, of which a proportion of P_0 are alive at the start of the next projection interval. This measure is named *fecundity rate* (symbol: F_x) (Caswell 2001). These "principles" can be written mathematically as:

$$N_{x,t+1} = P_{x-1}N_{x-1,t} \quad x > 0 \tag{11.2a}$$

$$N_{0,t+1} = \sum_{0}^{\omega} F_x N_{x,t} \tag{11.2b}$$

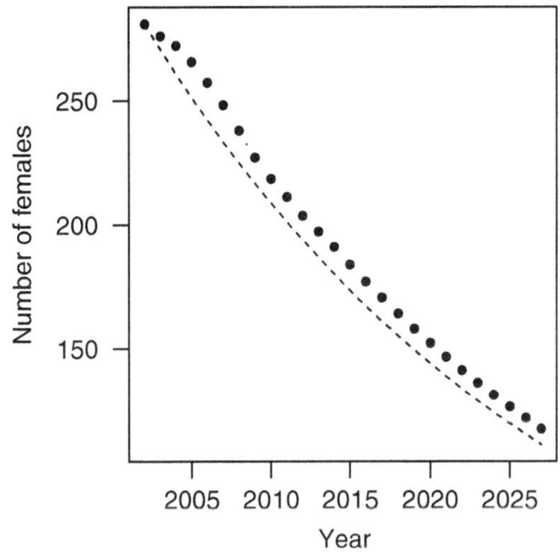

Fig. 11.3 Projection of the female snow leopard population in northern hemisphere zoos. The *solid dots* represent the results of a Leslie matrix projection; the *dashed line* represents the projected population size based on $\lambda = 0.96$ as estimated from life table data

where N_x is the number of individuals in age class x; t and $t + 1$ are intervals; ω is the oldest age class. Note that the initial age class is 0, conforming to the original paper by P.H. Leslie (1945).

The fecundity rate (m_x), which refers to mean number of offspring per individual (of the same sex), in fecundity tables (Chap. 7) is named *maternity function* within the context of Leslie matrices (Caswell 2001). This convention is followed in this chapter in order to avoid confusion between F_x and m_x.

The exact methods used to estimate F_x and P_x depend on the type of birth season i.e. *birth–flow* or *birth–pulse*; and the timing of census i.e. *pre–breeding* or *post–breeding* for birth–pulse reproduction. Details of these methods and their use in the Leslie matrix will be discussed in Sects. 11.3.1 and 11.4, respectively.

Source code for matrix population models that were originally written in MATLAB (Caswell 2001) have been "translated" to R script. The R package *popbio* (Stubben and Milligan 2007) is used in this chapter.

Figure 11.3 shows the Leslie matrix projection of the female snow leopard population in northern hemisphere zoo regions. The dashed line refers to a projection based on the annual growth rate (λ) estimated from the life table data without taking the age distribution into account. The northern hemisphere subset of the studbook data has been used, as the breeding season differs between north and south. This allows the selection of a birth–pulse (post–breeding) model for this species. Life table data estimated over the period 1983–2002 for this subpopulation are similar to those presented in Tables 7.3 and 7.4. The age distribution on 30 August 2002 (post–breeding) is used.

The female population is declining by about 4 % a year, i.e. $\lambda = 0.96$. However, the Leslie matrix projection shows the decline is less than expected from λ during the first 10 years. After that period of time the population will gradually decrease, conforming to the estimated λ.

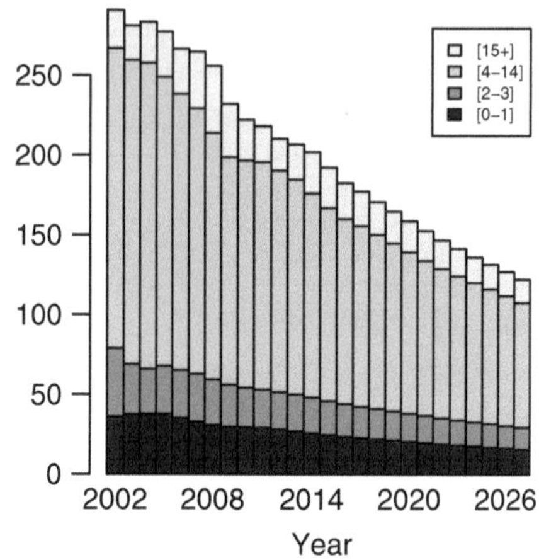

Fig. 11.4 Leslie matrix projection of female snow leopards in the northern hemisphere zoo region. The Y–axis represents the total population size. The *coloured bars* refer to age groups that correspond with life stages

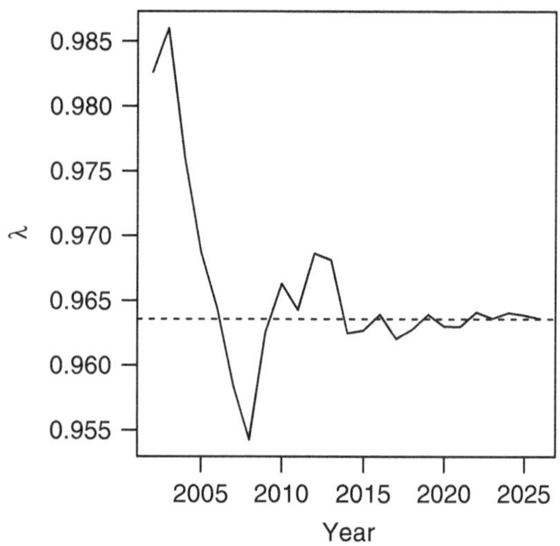

Fig. 11.5 Damping in annual population growth (λ) of female snow leopard population. The *dotted line* refers to $\lambda = 0.964$

A better insight into these different phases of population development can be obtained by including the individual age classes (see Fig. 11.4). Plotting some 20 classes results in a "cluttered" image that hardly allows the detection of trends. Therefore, age classes have been grouped in juvenile (0–1), subadult (2–3), adult (4–14) and senescence (15+) stages.

The number of births initially increases as juveniles and subadults become sexually mature. However, reproduction is not sufficient to replace the adult and the senescence groups, and population decline is faster until convergence to a stable age distribution (see Fig 9.2) has been reached. The population then declines as expected from λ. This effect is named *damping*.

Figure 11.5 shows damping in annual population growth (λ) in female snow leopards. Converging of λ starts, as would be expected from Figs. 11.3 and 11.4, around year 10. This *convergence time* conforms to the value (10.1 years) as calculated from the Leslie matrix.

Information about convergence time is important for population management because measures that change population growth do not necessarily have an immediate effect.

11.3.1 Birth–Pulse

Birth–pulse means that all births of the season occur at the same time. The timing of census therefore affects which age classes can be counted.

11.3.1.1 Post–breeding

A post–breeding census immediately after the birth–pulse will include all newborn individuals at the start of age class 0. The survival rate P_x in this case is equal to the p_x in the life table (Chap 7):

$$P_x = l_{x+1}/l_x \tag{11.3}$$

where l_x and l_{x+1} are survivorship values for classes x and $x + 1$, respectively.

The earliest breeding season after a post–breeding census is just before the next census. This means that individuals must have survived an entire interval before reproducing. The fecundity rate (F_x) in post–breeding is calculated as:

$$F_x = P_x m_x \tag{11.4}$$

where P_x is the survival probability and m_x is the maternity function for age class x.

11.3.1.2 Pre–breeding

The age distribution at pre–breeding census starts with survivors of the previous birth season that have reached class 1. These individuals become the first age class in the projection matrix, i.e. P_0 is the survival rate of individuals in the age interval 1–2 in this model. The survival rate is then calculated as:

$$P_x = l_{x+2}/l_{x+1} \tag{11.5}$$

where l_{x+1} and l_{x+2} are survivorship values for classes $x+1$ and $x+2$, respectively.

Since breeding takes place (immediately) after the pre–breeding census, only offspring that have survived to the first census will be counted in the fecundity rate of parents in class (F_x):

$$F_x = l_0 m_x \tag{11.6}$$

where l_0 is the survival rate of newborns and m_x is the maternity function for age class x.

11.3.2 Birth–Flow

Individuals in birth–flow reproduction can be born at any time within a projection interval. This means individuals belonging to the same age class have different ages. Unfortunately this information is "lost" in an age–structured population. The solution is to approximate P_x by calculating its midpoint value:

$$P_x \approx L_{x+1}/L_x \tag{11.7}$$

where L_{x+1} and L_x are midpoint survivorship values for classes $x + 1$ and x, respectively. See Eq. 7.7 for estimation of L_x.

The fecundity rate (F_x) is approximated in a similar way:

$$F_x \approx L_0 \left(\frac{m_x + P_x m_{x+1}}{2} \right) \tag{11.8}$$

where L_0 is the midpoint survivorship of age class 0, P_x is the midpoint survival rate, and m_x and m_{x+1} fecundity rates in classes x and $x + 1$.

11.4 Some Matrix Algebra

Matrix algebra is the original methodology for population projections based on age–specific fecundity and mortality data i.e. the Leslie matrix (Leslie 1945, 1948). Since studbook software such as *PMx* (Ballou et al. 2011) automate the process of constructing life tables and population projections, it is not a strict requirement to understand matrix algebra in detail. Moreover, software has also been developed for Leslie matrices, e.g. the R package *popbio* (Stubben and Milligan 2007).

There are clear advantages, though, in being familiar with matrix notations, as population matrix methods are used in ecological studies (which could involve "studbook" species). Equations 11.2(a) and (b) can be expressed as a matrix equation. The following example is for four age classes (with the first class numbered "0"):

$$\begin{bmatrix} N_0 \\ N_1 \\ N_2 \\ N_2 \end{bmatrix} (t+1) = \begin{bmatrix} F_0 & F_1 & F_2 & F_3 \\ P_0 & 0 & 0 & 0 \\ 0 & P_1 & 0 & 0 \\ 0 & 0 & P_2 & 0 \end{bmatrix} \times \begin{bmatrix} N_0 \\ N_1 \\ N_2 \\ N_3 \end{bmatrix} (t) \tag{11.9}$$

which can be concisely written as:

$$\mathbf{N}(t+1) = \mathbf{AN}(t) \tag{11.10}$$

where **A** is the projection matrix or Leslie matrix.

The top row of the matrix holds values for fecundity rates (F_x). Survival rates (P_x) are placed on the diagonal starting from row 2 of the first column. Growth rates, generation time, reproductive values and stable age distribution, as presented in Chap. 9, can be calculated from population matrices. In principle, a basic understanding of matrices, i.e. constructing a valid Leslie matrix, is required to use the available software.

11.5 Sensitivity and Elasticity

The results of population projections can indicate that management measures are required to (temporarily) increase or decrease the growth rate. Populations that reach carrying capacity require a decrease in growth rate (assuming that carrying capacity cannot be increased). Such a decrease is usually achieved by reducing reproduction. Populations that are declining or are growing too slowly, require measures that focus on increasing survival and/or reproduction.

Species' biology and knowledge of husbandry determine which measures to change population growth are feasible. For example, weaning age will determine the minimum inter–birth interval, and thus affects the number of females that can reproduce each breeding season. Husbandry may already have resulted in low (juvenile) mortality compared to the wild and/or past.

Measures to change population growth should not only be feasible to implement, but should be effective as well. One could experimentally test effects or *sensitivity* by manipulating individual fecundity and mortality rates in a table. Matrix calculus, however, offers methods to calculate the effect of a change in each individual rate in the matrix on λ.

This effect is expressed as the *sensitivity coefficient* (symbol: s_{ij}), which is the rate of increase in λ for an infinitesimal change of matrix element a_{ij}:

$$s_{ij} = \frac{\delta\lambda}{\delta a_{ij}} \tag{11.11}$$

where i and j are the row and column number of a matrix element, respectively (Caswell 2001).[1]

The result of sensitivity analysis is a matrix (S). Equation 11.12 presents a Leslie matrix (A) and a sensitivity matrix (S) for age classes 0 to 3 of female snow leopards in the northern hemisphere zoo region. The matrix cells for survival and fecundity (in italicised style), are the same as presented in Eq. 11.9.

Although only four age classes are shown, analysis involved 16 age classes, i.e. data are truncated after the last reproductive age class ($\lambda = 0.96$). Sensitivity

[1]It is easy to see the confusion with age classes starting at 0, whereas row and column numbers of a matrix start at 1.

analysis calculates the sensitivity coefficient for all matrix elements. Only matrix elements that represent survival or fecundity rates are of interest for population growth. These are indicated in bold font.

$$A = \begin{bmatrix} \mathit{0} & \boldsymbol{\mathit{0}} & \boldsymbol{\mathit{0.011}} & \boldsymbol{\mathit{0.156}} \\ \mathbf{0.686} & 0 & 0 & 0 \\ 0 & \mathbf{0.981} & 0 & 0 \\ 0 & 0 & \mathbf{0.985} & 0 \end{bmatrix} \quad S = \begin{bmatrix} \mathit{0.118} & \boldsymbol{\mathit{0.084}} & \boldsymbol{\mathit{0.085}} & \boldsymbol{\mathit{0.087}} \\ \mathbf{0.165} & 0.118 & 0.120 & 0.122 \\ 0.162 & \mathbf{0.115} & 0.118 & 0.120 \\ 0.157 & 0.112 & 0.114 & 0.117 \end{bmatrix} \tag{11.12}$$

Large sensitivity coefficients have more effect on λ than small ones. Matrix S shows that population growth is more sensitive to mortality changes in age class 0 than to age class 1, i.e. sensitivity coefficients s_{21} and s_{32} are 0.165 and 0.115, respectively. The survival rate of newborns (P_0) is 0.686. The effect of increasing the survival rate P_0 with 0.1 on λ can be calculated as follows:

$$\Delta\lambda = s_{21} \times \Delta P_0 = 0.165 * 0.1 = 0.0165 \tag{11.13a}$$

$$\lambda = 0.726 + 0.0165 = 0.7425 \tag{11.13b}$$

Sensitivity coefficients for survival rates are higher than those for fecundity rates (0.084 and 0.085 in matrix S). However, effects between these groups of rates can be difficult to compare as the range for survival rates is bound by the range 0 to 1, while fecundity rates can be larger than 1. A solution to this problem is to calculate the proportional effect, named *elasticity* (symbol: e_{ij}):

$$e_{ij} = \frac{a_{ij}}{\lambda} s_{ij} \tag{11.14}$$

where a_{ij} is a Leslie matrix element and s_{ij} a sensitivity coefficient, at row i and column j.

$$A = \begin{bmatrix} \mathit{0} & \boldsymbol{\mathit{0}} & \boldsymbol{\mathit{0.011}} & \boldsymbol{\mathit{0.156}} \\ \mathbf{0.686} & 0 & 0 & 0 \\ 0 & \mathbf{0.981} & 0 & 0 \\ 0 & 0 & \mathbf{0.985} & 0 \end{bmatrix} \quad E = \begin{bmatrix} \mathit{0} & \boldsymbol{\mathit{0.001}} & \boldsymbol{\mathit{0.007}} & \boldsymbol{\mathit{0.014}} \\ \mathbf{0.117} & 0 & 0 & 0 \\ 0 & \mathbf{0.117} & 0 & 0 \\ 0 & 0 & \mathbf{0.117} & 0 \end{bmatrix} \tag{11.15}$$

Equation 11.15 presents the results of elasticity analysis of matrix A in a new matrix E. The elasticity coefficients e_{21}, e_{32} and e_{43} of survival in classes 0, 1 and 2, respectively, are 0.117. This means that changes to these survival rates will have the same proportional effect on λ. Elasticity coefficients for young reproductive age classes are very low (i.e. 0.001, 0.007 and 0.014). This means that improving fecundity in the young age classes has hardly any effect on λ.

Sensitivity analysis can be used in demographic management to determine how an increase in population growth can be achieved effectively. In the example of snow leopards, the focus would be to improve survival in the young age classes. However, sensitivity analysis does not tell us whether measures are feasible.

For example, the mean (and median) litter size in snow leopards is 2. This means that the fecundity rate is expected to be 2 when all females in an age class reproduce. Increasing neonatal and juvenile survival rates may require human interventions, such as hand–rearing, that may affect socialisation and behaviour that are not necessarily beneficial for species to be reintroduced or for future reproduction.

11.6 Simulations

Fecundity and mortality are subject to *demographic stochasticity* due to the sampling process (Shaffer 1981). The results of bootstrap techniques/analyses on fecundity and mortality rates in Chap. 8 illustrate this type of uncertainty in population growth.

Demographic stochasticity is one of the four broader groups of uncertainties (*environmental stochasticity*, *natural catastrophes* and *genetic stochasticity*) that are considered in species conservation (Shaffer 1981). These four groups of uncertainties can be modelled for studbook populations with the software program VORTEX (Lacy and Pollak 2014; Lacy et al. 2014).

Projection models of studbook populations include demographic stochasticity e.g. ZooRisk (Earnhardt et al. 2008) and PMx. The R package *popbio* (Stubben and Milligan 2007) also supports demographic stochasticity through simulation and is used in this section. This software assumes a binomial and Poisson distribution for mortality and fecundity, respectively.

Figure 11.6 presents the results of a stochastic Leslie matrix model with 100 runs for the female snow leopard population. The 95 % projection interval for year 2027 is between 89 and 169 females. This wide interval illustrates that demographic stochasticity needs to be taken into account in the management of small populations.

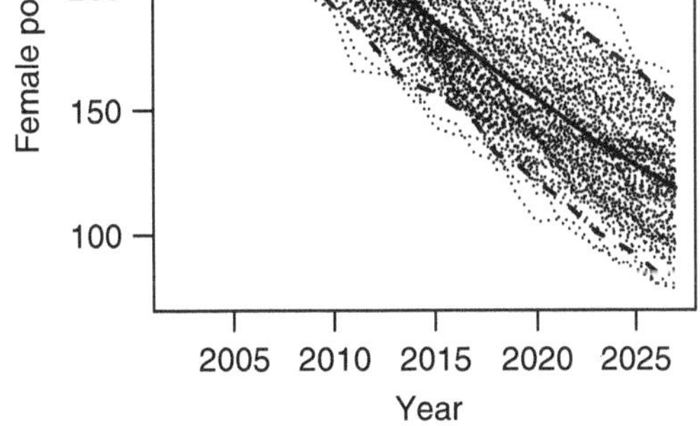

Fig. 11.6 Stochastic population projection of female snow leopard population (100 runs). *Dotted lines* are individual runs; the *solid line* is the mean female population size. *Dashed lines* represent the 2.5 and 97.5 percentiles

The Poisson distribution is not necessarily the optimal random model for fecundity (see Chap. 8). Nevertheless, even simple stochastic models show how population development can deviate from deterministic models that are based on fixed growth rates.

11.7 Projection or Prediction

Population projections as presented above are often interpreted as forecasts or predictions. However, this interpretation is not correct. Projection refers to what *would* happen with the population *if* the current fecundity and mortality schemes do not change. Forecast or prediction refers to what *will* happen to a population with a certain probability (Caswell 2001; Keyfitz 1972).

Fecundity and survival rates are not likely to stay constant in the future. Management measures to reduce breeding will be taken whenever the population is growing towards filling the available space (carrying capacity), e.g. logistic growth in the European blesbok population (Fig. 11.2). Husbandry continuously improves and can result in lower mortality and/or higher fecundity. This particularly applies to species that were not maintained in captivity until recently, e.g. species that are rescued as the response to the amphibian crisis caused by the chytrid fungus (Cheng et al. 2011) and global warming (climate change) (Raxworthy and Nussbaum 1996). Inbreeding is another factor that can (negatively) affect fecundity and mortality over time (see Chap. 13).

The simulation version of the Leslie matrix as presented in the previous section provides percentiles in expected population size (Fig 11.6). This may wrongly suggest that this stochastic model provides a prediction. The result, however, is still a projection as only demographic stochasticity is considered. Software programs that model the major categories of uncertainties, e.g. VORTEX (Lacy and Pollak 2014; Lacy et al. 2014), come close to predicting population development.

11.8 Back to the Future

Stochastic models are useful, whether qualified as prediction or projection, to give an idea of the potential range of future population size. More interesting, however, is to know whether the selected model parameters are correct, and hence whether the real population size will be within the projected range. Part of this knowledge can be achieved through retrospective studies in which projections of the population in the past are compared with census data (e.g. Brook et al. 1997).

Figure 11.7a shows a logistic growth model that includes a projection for the period 1995–2001 for the 1995 female snow leopard population. The logistic growth model matches 1985–1995 census data ($\chi^2 = 1.76, p = 1.0, df = 12$). The population is projected, retrospectively, to grow between 1995 and 2001 up to a

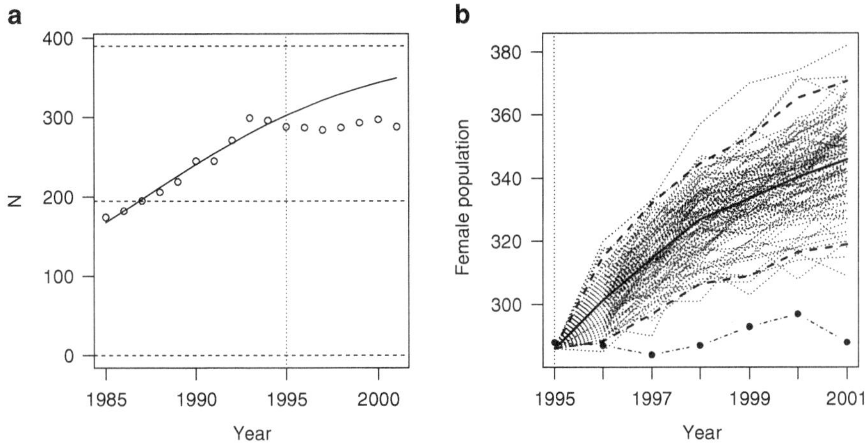

Fig. 11.7 (**a**) Logistic growth projection of the 1995 female snow leopard population (*solid line*) based on 1985–1995 census data. *Open circles* are actual census data. See text for details. (**b**) Stochastic matrix projection based on 1985–1995 life tables. *Dotted lines* are individual runs; the line with *closed circles* represents actual post-breeding census data; the *solid line* is the mean projected female population size

carrying capacity of 390 (solid line). In reality, however, the female population slowly declined (open circles).

The stochastic matrix model based on life tables over the period 1985–1995 shows a similar pattern (Fig. 11.7b). Even simulation runs at the lower spectrum of the range are higher than the real census data. The discrepancy between projection and reality can be explained by using the 1985–1995 trend which resembles a period of linear growth. However, the female population was at a turning point in 1994 when it started to decline (Fig. 11.7a). Annual growth rates are 1.0 and 0.97 for the years 1994 and 1995, respectively.

These opposite trends illustrate a dilemma in selecting the correct time period. Growth rates derived from census analysis or life tables based on two years may have low validity. Yet, the results from this retrospective analysis show that the 1994–1995 trend actually would have been a better indicator for population growth. Life tables based on imaginary cohorts are, in effect, averages, and consequently the growth rates calculated from these tables are averages. In this example, the average λ is larger than 1.

Census–like analyses such as crude birth and crude death rates are useful as indicators of changing trends. For example, the annual birth rate in snow leopard females in the global studbook population was declining within the 1985–1995 period (Fig. 9.3b). This trend was the result of a management measure to reduce population growth by keeping offspring with their mother longer (L. Blomqvist, personal communication). This measure is reflected in the median inter–birth interval which increased from 423 days in the period 1974–1984 ($n = 79$) to 717 days in the period 1985–1995 ($n = 111$).

Population projections assume that mean fecundity and mortality rates do not change during the projected period of time. The above example shows that management measures can drastically change these vital rates in order to control population size. There is no real remedy, other than adapting population projections based on real population size on a regular basis.

11.9 Stage–Based Models

Life tables of studbook populations are age–based. This model is meaningful for species where individual survival and reproduction (*vital rates*) are age related. For example, snow leopards reach sexual maturity at the age of two years in captivity (see Table 7.4). However, vital rates are not necessarily directly related to age, but can depend on environmental and/or social conditions (e.g. Caswell 2001).

Ambient temperature and/or food quality can have impact on somatic growth in ectotherms, resulting in individuals reaching sexual maturity at different ages, e.g. Atlantic salmon (*Salmo salar*) (Jonsson et al. 2013). A study on captive green turtles (*Chelonia mydas*) revealed the difficulty of estimating age at sexual maturity, but showed that carapace length seems to approach a threshold for maturity (Bjorndal et al. 2013); thus carapace length, instead of age, needs to be used to determine whether a green turtle is (likely to be) sexually mature.

Pheromonal signals of dominant females in common marmosets (*Callithrix jacchus*) can suppress ovulation of subordinate females (daughters) (Barrett et al. 1990). Reproductive suppression has been observed in various social mammalian species (Beehner and Lu 2013; Clutton–Brock and Huchard 2013; Wasser and Barash 1983). This means that age–specific fecundity rates are also determined by social settings.

Stage–based models are more appropriate for population projections of species in which vital rates are largely determined by external conditions (Caswell 2001). Common categories such as neonate, juvenile, subadult and adult in mammals (Kunz et al. 1996) or eggs, larvae, pupae and adults in insects (Lefkovitch 1965), can be used in stage models. Stage categories are not limited to developmental stages. Other categories such as reproductive status or management category can be used as well (Faust et al. 2003). Stage–models can also be an alternative to age–based models when the exact age of individuals can not be assessed (Lefkovitch 1965).

When age is not a reliable indicator of the developmental stage (state) of individuals, other methods are required. Physical features, such as carapace length in marine turtles (Crouse et al. 1987) and tortoises (Doak et al. 1994), or the peaked sagittal crest in adult silverback gorillas (*Gorilla* spp.) (Breuer et al. 2009), are examples of markers that can be used to identify stages. Behavioural features can also be used, e.g. weaning to identify the start of the juvenile stage (Kunz et al. 1996).

Projections of stage–structured populations are similar to those described for age–structured populations. The *Lefkovitch matrix* (Lefkovitch 1965) is a generalisation of the Leslie matrix (Leslie 1945, 1948) that was developed for stage–based models. Major differences with age–based models are that stages can differ in their duration and that individuals may remain in the same stage during a projection interval (Crouse et al. 1987). This means that the probabilities of surviving to the next stage and of then entering the next stage, during a single time step, are required. The symbols P_i for survival and remaining in stage i, and G_i for survival and entering the next stage $(i + 1)$, are commonly used.

Stage–based models have not been implemented, so far, in studbook management software. The reason(s) may be historical. The first managed programmes involved species such as the red panda and the snow leopard, whose life cycles can be described by an age–based model. Furthermore, the strength of studbooks is the availability of data on dates of birth and death, i.e. age, which makes it relatively easy to automate construction of age–based life tables.

Practical reasons for studbook management may favour stage over age–based models. Managers are more likely to think in terms of numbers of infants or adults than in numbers of one–year–olds or eight–year–olds. This does not necessarily imply that stage–based models need to be used. For example, age distributions can be presented as stage distributions by grouping age classes (see Fig. 11.4).

Grouping age classes into biological stages has the advantage of reducing problems associated with small sample sizes (Faust et al. 2003). Age–class models with flexible class widths could be implemented in studbook software to facilitate the use of stages. Since the stage of an individual is directly related to its age, it should be relatively easy to implement "age–stage" models in age distribution and life table analysis.

Stage–based population projections require information on the *transition probability* (G_i) that an individual which survives, changes stage during a projection interval (e.g. one year). This probability is clear when stages are related to age, i.e. $G_i = 0$ when an individual is too young, and $G_i = 1$ (100 %) when an individual has reached the age of the next stage.

Each individual within a stage, regardless of its age, has the same probability of surviving (P_i) during a projection interval. However, the Leslie matrix does not "follow" individuals. This means that ages of surviving individuals at the end of a projection interval, and thus the transition to the next stage, are not "known". Matrix projections require making assumptions regarding the age distribution and survival *within* each stage (Caswell 2001; Crouse et al. 1987).

The transition probability adds an extra (stochastic) uncertainty to population projections when applying stage models to age–structured populations in matrix calculus. This uncertainty can be avoided by "tracking" individuals in simulation models that implement pedigrees, e.g. VORTEX (Lacy and Pollak 2014).

Species' biology and available data determine which model is most appropriate for demography and for population projections in particular. Cooperative breeding programmes in zoos include taxa for which the life–cycle may be better described by stage than by age structure e.g. crocodiles and tortoises (European Association of Zoos and Aquaria 2016). Furthermore, these stage models are useful for including management measures that are not (completely) age dependent e.g. use of contraceptives. This requires that stage transitions are registered in studbooks (similar to periods of contraception in SPARKS (Scobie and Bingaman Lackey 1997–2012)), and that studbook software can handle stage–based models.

References

Ballou JD, Lacy RC, Pollak JP (2011) PMx: software for demographic and genetic analysis and management of pedigreed populations (version 1.0). Technical report Chicago Zoological Society, Brookfield

Barrett J, Abbott DH, George LM (1990) Extension of reproductive suppression by pheromonal cues in subordinate female marmoset monkeys, *Callithrix jacchus*. J Reprod Fertil 90(2):411–418

Beehner JC, Lu A (2013) Reproductive suppression in female primates: a review. Evolut Anthropol Issues News Rev 22(5):226–238

Bjorndal K, Parsons J, Mustin W, Bolten A (2013) Threshold to maturity in a long–lived reptile: interactions of age, size, and growth. Mar Biol 160(3):607–616

Breuer T, Breuer-Ndoundou Hockemba M, Olejniczak C, Parnell RJ, Stokes EJ (2009) Physical maturation, life history classes and age estimates of free–ranging western gorillas – insights from Mbeli Bai, Republic of Congo. Am J Primatol 71:106–119

Brook BW, Lim L, Harden R, Frankham R (1997) Does population viability analysis software predict the behaviour of real populations? A retrospective study on the Lord Howe Island woodhen *Tricholimnas sylvestris* (Sclater). Biol Conserv 82(2):119–128

Caswell H (2001) Matrix populations models: construction, analysis, and interpretation, 2nd edn. Sinnauer Associates, Sunderland

Cheng TL, Rovito SM, Wake DB, Vredenburg VT (2011) Coincident mass extirpation of neotropical amphibians with the emergence of the infectious fungal pathogen *Batrachochytrium dendrobatidis*. Proc Natl Acad of Sci United States Am 108(23):9502–9507

Clutton–Brock TH, Huchard E (2013) Social competition and selection in males and females. Philos Trans R Soc Lond B Biol Sci 368(1631). doi:10.1098/rstb.2013.0074

Crouse DT, Crowder LB, Caswell H (1987) A stage-based population model for loggerhead sea turtles and implications for conservation. Ecology 68(5):1412–1423

Doak D, Kareiva P, Klepetka B (1994) Modeling population viability for the desert tortoise in the Western Mojave Desert. Ecol Appli 4(3):446–460

Earnhardt JM, Bergstrom YM, Lin A, Schloss CA, Thompson SD (2008) ZooRisk: a risk assessment tool version 3.8. Lincoln Park Zoo, Chicago

European Association of Zoos and Aquaria (2016) EAZA breeding programme overview. http://www.eaza.net/assets/Uploads/CCC/Overview-EAZA-Breeding-Programmes-July-2016.pdf

Faust LJ, Thompson SD, Earnhardt JM, Brown E, Ryan S, Sherman M, Yurenka M (2003) Using stage-based system dynamics modeling for demographic management of captive populations. Zoo Biol 22:45–64

Jonsson B, Jonsson N, Finstad AG, Genner M (2013) Effects of temperature and food quality on age and size at maturity in ectotherms: an experimental test with Atlantic salmon. J Anim Ecol 82(1):201–210

Keyfitz N (1972) On future population. J Am Stat Assoc 67(338):347–363

Kunz TH, Wemmer C, Hayssen V (1996) Sex, age, and reproductive condition of mammals. In: Wilson DE, Cole FR, Nichols JD, Rudran R, Foster MS (eds) Measuring and monitoring biological diversity: standard methods for mammals. Smithsonian Institution Press, Washington DC, pp 279–290

Lacy RC, Pollak JP (2014) VORTEX: a stochastic simulation of the extinction process, version 10.0. Technical report, Chicago Zoological Society, Brookfield
Lacy RC, Miller PS, Traylor–Holzer K (2014) VORTEX 10 user's manual. IUCN SSC Conservation Breeding Specialist Group, Apple Valley
Lefkovitch LP (1965) The study of population growth in organisms grouped in stages. Biometrics 21(1):1–18
Leslie PH (1945) On the use of matrices in certain population mathematics. Biometrika 33(3):183–212
Leslie PH (1948) Some further notes on the use of matrices in population mathematics. Biometrika 35(3/4):213–245
Norris N (1940) The standard errors of the geometric and harmonic means and their application to index numbers. Ann Math Stat 11(4):445–448
Princée FPG (1998) Genetic management of small animal populations in zoos and wildlife reserves: the use of genome models in the estimation of genetic variation and the effects of social structures. PhD thesis, University of Groningen, Groningen
Raxworthy CJ, Nussbaum RA (1996) Montane amphibian and reptile communities in Madagascar. Conserv Biol 10(3):750–756
Scobie P, Bingaman Lackey L (1997–2012) Single Population Animal Record Keeping System (SPARKS) Versions 1.42–1.66. International Species Information System, Eagan
Shaffer ML (1981) Minimum population sizes for species conservation. BioScience 31(2):131–134
Stubben C, Milligan B (2007) Estimating and analyzing demographic models using the popbio package in R. J Stat Software 22(11):1–23
Wasser SK, Barash DP (1983) Reproductive suppression among female mammals: implications for biomedicine and sexual selection theory. Q Rev Biol 58(4):513–538

Part III
Genetic Analyses

Chapter 12
Genetic Variation and Generations

Abstract Maintenance of the original (wild) genetic variation is an important objective in the management of small populations of wild species. Several measures to quantify genetic variation exist. These measures are polymorphism or proportion of polymorphic loci, average number of allelic variants per locus, observed heterozygosity and average expected heterozygosity or gene diversity. Methods to calculate these measures are presented. An introduction to genetic loss in small populations due to consanguineous matings (inbreeding), genetic drift and selection (intentional and unintentional) is provided. Genetic loss is expressed as loss per generation. It is common practice to assign generation 0 to the founder group. There are various rules to calculate generation number when generations overlap. The rule used by the Convention on International Trade in Endangered Species of Wild Fauna and Flora (CITES) and rules used in different fields of species management are described. These include maternal generation, genetic generation and average generation. A hypothetical pedigree example is used to illustrate differences between these rules. Long–term effects of the captive environment on species characteristics can be measured by comparing generation groups. The season of birth in maternal generation groups within Nepalese red pandas in the northern hemisphere is used as an example. The pairwise Wilcoxon rank test is applied to test for differences between seven generations.

12.1 Introduction

Genetic variation is considered an important requirement for both short–term and long–term survival of species, as it provides the "resource" to cope with changing environments. The negative effects of inbreeding and the importance of genetic variation in zoo populations were recognised in the late 1970s (e.g. Flesness 1977; Ralls et al. 1979). This changed studbooks from being merely a register into cooperative management programmes in which population genetics can be considered the "backbone". Maintenance of genetic variation and avoidance of inbreeding have become important objectives in management of small populations (e.g. Lacy et al. 1995; Princée 1998). Several measures to quantify genetic variation are used in the management of captive populations (e.g. Lacy 1995). The next section provides a short overview of the most common ones.

F.P.G. Princée, *Exploring Studbooks for Wildlife Management and Conservation*, Topics in Biodiversity and Conservation 17, DOI 10.1007/978-3-319-50032-4_12

12.2 Genetic Measures

A relatively crude measure of genetic variation is *polymorphism* (symbol: P), which is the proportion of polymorphic loci in a population:

$$P = 1 - \frac{r_p}{r} \tag{12.1}$$

where r is the number of loci, and r_p is the number of polymorphic loci (Nei et al. 1975).

Technically speaking, any locus with two or more allelic variants is polymorphic. However, a rare allele may not be observed in a population sample. Therefore, a more practical definition – a frequency of less than 99 % of the most common allele in the population – is generally applied (Nei et al. 1975). This frequency is arbitrary and other levels, e.g. 95 %, can be found in the literature.

The *average number of alleles per locus* (symbol: $\overline{n}$) in a population provides more detail on genetic variation than polymorphism:

$$\overline{n} = \frac{1}{r}\sum_{j=1}^{r} n_j \tag{12.2}$$

where r is the number of loci and n_j the number of (observed) alleles at the jth locus.

Polymorphism and (average) number of alleles do not provide complete information on genetic variation in diploid species, as the heterozygous or homozygous state of individuals is not considered. Two (slightly) different measures of *heterozygosity* are used.

The *observed heterozygosity* (symbol: H_o) is estimated from observed heterozygous loci in individuals:

$$H_o = \frac{\sum_{x=1}^{N} h_x}{Nr} \tag{12.3}$$

where N is the number of individuals, r is the number of loci and h_x is the number of heterozygous loci of individual x (Nei et al. 1975).

The *average expected heterozygosity* (symbol: H_e) estimates the expected proportion of heterozygous loci from allele frequencies.

The expected heterozygosity at a single locus (h_e) is estimated as:

$$h_e = 1 - \sum_{i=1}^{n} p_i^2 \tag{12.4}$$

where n is the number of alleles and p_i is the frequency of the ith allele (Nei et al. 1975).

The average expected heterozygosity (H_e) is estimated as:

$$H_e = \frac{\sum_{j=1}^{r} h_{ej}}{r} \tag{12.5}$$

where r is the number of loci and h_{ej} is the gene diversity at the jth locus (Nei et al. 1975).

In the literature, the term heterozygosity is often used without the adjectives observed or expected. Both measures have the same value under random mating conditions, but can differ under inbreeding and/or assortative mating (see Chap. 14). The alternative name *gene diversity* is preferred when referring to heterozygosity based on allele frequencies.

Small populations are affected by demographic and genetic processes that can result in loss of genetic variation, i.e. increase in homozygosity and loss of allelic variants. Populations of endangered species, whether in captivity or in the wild, are in general small in size and, therefore, subject to genetic loss. Three major causes of genetic loss can be identified:

1. Inbreeding, due to matings between related individuals (*consanguineous matings*), which results in increased homozygosity and expression of recessive deleterious alleles.
2. Genetic drift, a random change in allele frequencies from generation to generation that results in increased homozygosity and loss of allelic variation.
3. Selection, either intentional or unintentional, whether in favour of or against specific genotypes and phenotypes, results a priori in loss of genetic variation.

The next chapters in the genetic section present and explore techniques to detect inbreeding depression and to estimate genetic loss in studbook populations. This section also explores the use of life history traits that can be extracted directly from studbooks for quantitative genetic studies and detection of (phenotypic) selection.

12.3 Generation

Genetic loss in populations is generally expressed in terms of loss in genetic variation per generation. But what is a generation? Generation refers to "in a line of descent, individuals that share a common ancestor and are all the same number of broods away from that ancestor" (Lawrence 2008). The "ancestor" in captive populations refers to a wild–caught individual which reproduced (*founder*).

The method for calculating generation (number) seems to be relatively easy. Offspring have a generation number of one higher than their parents. Thus, offspring of, for example, wild–caught parents (P_0), belong to the first generation (F_1). Offspring of F_1 parents (also denoted as P_1) belong to generation F_2, and so on. This seems straightforward, yet it ignores the fact that generations can overlap. Generation overlap can occur in any species with a reproductive lifespan that is longer than the age at sexual maturity. This means that both parental and new generation groups are reproductive and, thus, can breed together.

Zoo populations can show extreme generation overlap (see e.g. Princée 1988). First, many zoo populations have a long history, sometimes going back to the early 1900s, e.g. European bison (Princée 1998), where some zoos were successful in breeding, and other zoos depended on import from the wild. Second, reproductive lifespan of individuals in zoos can be extended, compared to the wild, as a result of stable food resources, veterinary care and the absence of predation, thus increasing the chances that generation overlap occurs.

Different rules for computing generation number exist (Brinks et al. 1961; Princée 1988; Schwitzer and Kaumanns 2009; Wijnstekers 2011). They will be discussed in the next section.

12.3.1 Calculating Generations

The *Convention on International Trade in Endangered Species of Wild Fauna and Flora (CITES)* requires the generation number of individuals to determine whether, or under which conditions, permits for import and export will be issued. CITES applies the so-called "shortest link" in line of descent to calculate the generation number of captive-born offspring (Wijnstekers 2011). The two major rules in the CITES approach – Resolution Conf. 10.16 (Rev.) – are:

> "first–generation offspring (F1)" are specimens produced in a controlled environment from parents at least one of which was conceived in or taken from the wild;
> "offspring of second generation (F2) or subsequent generation (F3, F4, etc.)" are specimens produced in a controlled environment from parents that were also produced in a controlled environment;

This means that an individual which has one wild–caught parent is, according to CITES, also considered to be F_1, regardless of the generation number of the other parent. The calculation rule in cases of generation overlap beyond the F_2 generation is described in the CITES Resolution 10.16 (rev) (Wijnstekers 2011). However, in general the "shortest link" is also applied in these cases. For example, offspring of an $F_1 \times F_9$ mating will be assigned F_2 (C. Schürmann, personal communication).

The implications of this rule for obtaining permits is beyond the scope of this book. However, generation numbers calculated according to CITES may not be biologically sound when used in studies to evaluate the environmental effects of captivity, such as shifts in birth season between generation groups. It would be better to calculate generation number of captive–born offspring following the *maternal line of descent* (which ignores the generation of the male) (Schwitzer and Kaumanns 2009). The generation number of offspring is the generation of the dam plus one.

Note that the "maternal" rule should not be used in population genetic studies where genetic variation in captive generation groups is to be assessed, as the male contribution to generation number is ignored.

Table 12.1 Generation number of captive–born individuals in the pedigree example (Fig. 12.1) according to different calculation rules

Individual	F_{CITES}	$F_{maternal}$	$F_{genetic}$	$\bar{F}$
7	1	1	1	1
8	1	1	1	1
9	2	2	2	2
10	1	3	3	2
11	1	1	4	2

F_{CITES} CITES generation number, $F_{maternal}$ maternal generation number, $F_{genetic}$ "genetic" generation number, $\bar{F}$ average generation number

A third rule, *genetic generation*, based on the "longest link of descent" is used in the genetic simulation model GeneFlow (Princée 1988). Offspring are assigned a generation number one higher than the parent with the highest generation number. This means that offspring have by definition a higher generation number than their parents.

A fourth rule, *average generation* (Brinks et al. 1961), is recommended for use when generations overlap (Calboli et al. 2008). Average generation is calculated according to the following "rule":

$$\bar{F}_{offspring} = \frac{\bar{F}_{sire} + \bar{F}_{dam}}{2} + 1 \tag{12.6}$$

Generation numbers calculated from either CITES, genetic generation or average generation can be used in genetic studies that assess genetic variation in generation groups or genetic loss between generation groups. However, each rule can lead to different results when generations overlap, which need to be interpreted accordingly.

Table 12.1 summarises the differences between these rules based on the pedigree in Fig. 12.1. This hypothetical pedigree includes six wild–caught individuals (numbers 1 to 6), which consequently belong to the P_0 group. The differences between these rules appear in individuals 10 and 11. They both belong to the F_1 generation according CITES, as one of their parents is wild–caught. The difference between using "maternal" and "genetic" rules is illustrated with individual 11. Since the dam (6) is wild–caught, it belongs to the F_1 generation following the "maternal" rule and the F_4 generation following the "genetic" rule.

12.3.2 Effects of the Captive Environment: An Example

Effects of the captive environment on biological features, such as seasonality in birth, can be evaluated using generation groups. Generation numbers of individuals, especially based on the maternal rule (see Sect. 12.3.1), indicate the time–span a species – or better lineage – has been in captivity.

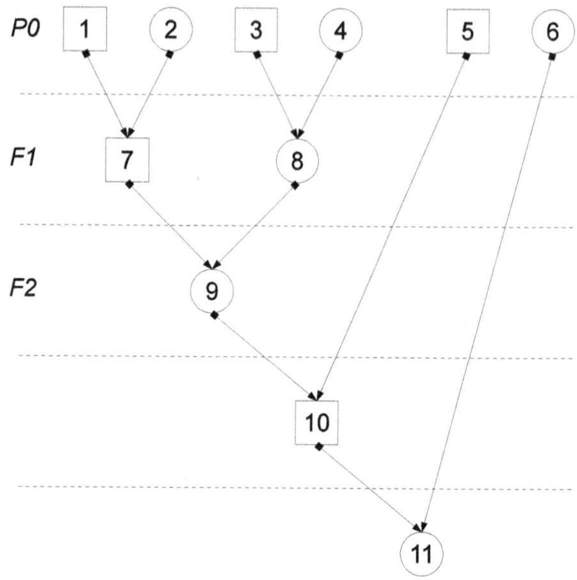

Fig. 12.1 Pedigree example of "genetic" generation groups. P_0 is founder group; F_1 and F_2 are captive generations
□ = *males*, ○ = *females*

Since the ancestors may have been imported and reproduced during different periods of time in the history of the captive population, individuals that were born in different periods of time can belong to the same generation overlap. This means that results of analyses on generation groups can be affected whenever husbandry changes over time. See for example differences in mortality in Sect. 3.2.4.

Seasonality in litter dates in Nepalese red pandas in the Northern hemisphere is used as an example to illustrate differences between generation groups, thus time in captivity. This species shows a strong seasonality in birth with 26 June as mean day (see Table 3.8). Since the birth season in the Northern hemisphere starts and ends within a calendar year, "conventional" statistical methods such as ANOVA and the Wilcoxon rank test can be applied (see Chap 19).

The time–span selected for this analysis includes the period that the international studbook was established (in 1977) so as to include births from wild–caught females. Generation groups were calculated according to the maternal rule (see Sect. 12.3.1).

Figure 12.2 shows box and whisker plots of seasonality in birth per generation group. Following the median (thick) lines in these boxes, one can see a tendency for a shift in mean day of parturition in females belonging to generations 3, 4 and 5. Mean day in generation 6 and 7 returns to the date of the early captive generations. However, statistical tests need to be carried out to determine whether the observed differences are significant.

Table 12.2 shows the results of an ANOVA on seasonality in birth and generation number. Since $p < 0.001$, it can be concluded that seasonality in birth differs significantly between generation groups. However, this statistic does not inform us which groups differ. This requires pairwise testing of each group. This could be the

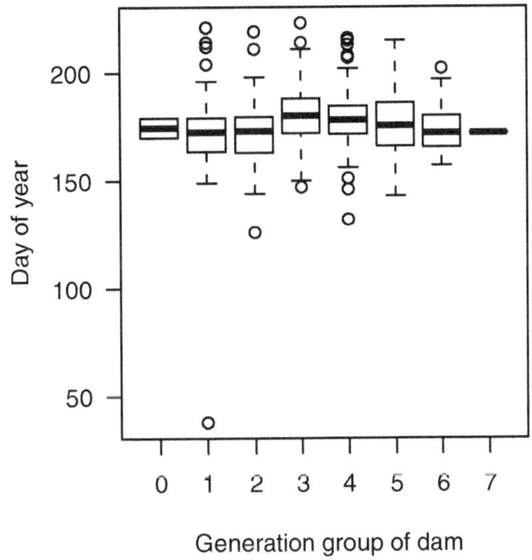

Fig. 12.2 Box–and–whisker plots showing shift in season of birth (litter date) in captive generations of Nepalese red pandas in the Northern hemisphere (period 1977–2007), $n = 559$. *Thick horizontal line*: median, bottom and top of box are first and third quartiles, "whiskers" are minimum/maximum (or roughly 2 standard deviations), *empty circles* are outliers

Table 12.2 ANOVA on seasonality in birth (litters) and captive generations of Nepalese red pandas in the Northern hemisphere (period 1977–2007). $n = 857$

	Df	SS	MS	F	$P_{[F]}$
Generation	7	9,109	1,301.28	5.9019	< 0.001
Residuals	849	187,191	220.48		

Df degrees of freedom, *SS* sum of squares, *MS* mean squares, *F* F–statistic, $P_{[F]}$ probability of F

Table 12.3 Pairwise Wilcoxon rank test (p–values) on seasonality in litter dates between founder and captive maternal generations (F) of Nepalese red pandas in the Northern hemisphere during the period 1977–2007 ($n = 857$)

F	0	1	2	3	4	5	6
1	1.00000						
2	1.00000	1.00000					
3	0.00432	0.00016	6.6e-05				
4	0.03419	0.00354	0.00362	1.00000			
5	1.00000	1.00000	1.00000	0.08031	0.46163		
6	1.00000	1.00000	1.00000	0.05477	0.20447	1.00000	
7	1.00000	1.00000	1.00000	1.00000	1.00000	1.00000	1.00000

pairwise t test or the non–parametric pairwise Wilcoxon rank test (R Core Team 2015). Since it is not known whether the birth data are distributed normally, it is safer to use the non–parametric test.

Table 12.3 presents the p–values for each generation pair, corrected with Holm's method for multiple comparisons as implemented in the R software (R Core Team 2015). Values of $p < 0.05$ indicate significant differences between groups.

Parturition date in generation groups F_3 and F_4 females differ significantly from those in groups F_0 (founders), F_1 and F_2.

Generation groups can also be used to study effects of time in captivity on differences in litter size and infant mortality (e.g Schwitzer and Kaumanns 2009). However, note that inbreeding can also affect infant mortality and may “interfere” with results for generation groups F_2 and higher. Inbreeding depression is discussed in the next chapter.

References

Brinks JS, Clark RT, Rice FJ (1961) Estimation of genetic trends in beef cattle (abstract). J Anim Sci 20:903

Calboli FC, Sampson J, Fretwell N, Balding DJ (2008) Population structure and inbreeding from pedigree analysis of purebred dogs. Genetics 179(1):593–601

Flesness NR (1977) Gene pool conservation and computer analysis. Int Zoo Yearb 17:77–81

Lacy RC (1995) Clarification of genetic terms and their use in the management of captive populations. Zoo Biol 14:565–578

Lacy RC, Ballou JD, Princée F, Starfield A, Thompson EA (1995) Pedigree analysis for population management. In: Ballou JD, Gilpin M, Foose TJ (eds) Population management for survival and recovery: analytical methods and strategies in small population conservation. Methods and cases in conservation science series. Columbia University Press, New York, pp 57–75

Lawrence E (ed) (2008) Henderson’s dictionary of biology, 14th edn. Pearson Education, Harlow

Nei M, Maruyama T, Chakraborty R (1975) The bottleneck effect and genetic variability in populations. Evolution 29(1):1–10

Princée FPG (1988) Genetic variation in the zoo population of the red panda subspecies *Ailurus fulgens fulgens*. Zoo Biol 7(3):219–231

Princée FPG (1998) Genetic management of small animal populations in zoos and wildlife reserves: the use of genome models in the estimation of genetic variation and the effects of social structures. PhD thesis, University of Groningen, Groningen

R Core Team (2015) R: a Language and Environment for Statistical Computing. R Foundation for Statistical Computing, Vienna. https://www.R-project.org/

Ralls K, Brugger K, Ballou J (1979) Inbreeding and juvenile mortality in small populations of ungulates. Science 206:1101–1103

Schwitzer C, Kaumanns W (2009) Litter size, infant mortality and female body weight in captive black–and–white ruffed lemurs *Varecia variegata*. Endanger Species Res 8:201–209

Wijnstekers W (2011) The evolution of CITES: a reference to the Convention on International Trade in Endangered Species of Wild Fauna and Flora, 9th edn. International Council for Game and Wildlife Conservation, Budapest

Chapter 13
Inbreeding

Abstract Studies at the end of the 1970s, on the negative effects of matings between related individuals (consanguineous matings) in zoo populations, can be considered as one of the main driving forces that resulted in introducing genetic management to cooperative breeding programmes. Calculation and interpretation of inbreeding coefficients is discussed. The main focus of this chapter is to detect inbreeding depression in the population directly from life history data in the studbook. The Nepalese red panda studbook population is used as an example. Methods to detect inbreeding depression in different stages of life are presented. Two–way contingency tables are used to test for differences in neonatal and juvenile mortality between inbred and non–inbred individuals. The Donner's test which adjusts for litter effect is applied. The Kaplan Meier product limit estimator and Cox proportional hazards regression models are used to detect differences in lifespan in inbred and non–inbred individuals. Different methods to estimate lethal equivalents are presented. These include linear and generalised linear regression, the generalised equation estimator (to adjust for pseudoreplication) and maximum likelihood estimation. Logrank tests are used to detect differences in age–specific fecundity between inbred and non–inbred individuals. The same test is also applied to age–specific unweighted reproductive values, which can be considered as a measure of fitness.

13.1 Introduction

Breeders of domestic animal and plant species knew about the negative effects of matings between related individuals (or *consanguineous matings*) long before inheritance and Mendelian laws were common knowledge.[1] For example, Darwin (1876) referred to the detrimental effects of consanguineous matings, such as reduced fecundity, in his study on variation in domestic animals and plants.

[1]Although Gregor Mendel published the results on inheritance in 1866, this work remained unknown until 1900 when Hugo de Vries published his work and referred in a footnote to Mendel's work.

F.P.G. Princée, *Exploring Studbooks for Wildlife Management and Conservation*, Topics in Biodiversity and Conservation 17, DOI 10.1007/978-3-319-50032-4_13

Mating between related individuals is known as *inbreeding* and its negative effect on fitness is known as *inbreeding depression* (symbol: δ). The result of inbreeding is an increase in homozygosity in offspring because they possess alleles that originate from the same ancestor i.e. *identical by descent* or *autozygous*. This increases the probability that deleterious alleles are expressed in the homozygous state.

This definition of the term inbreeding is generally used in the context of breeding animals and plants, whether domestic or wild (such as species in zoological gardens), for which pedigrees are available. However, different definitions of inbreeding exist which can lead to confusion (Allendorf et al. 2013; Keller and Waller 2002). These definitions will be discussed in Chap. 14. The definition as described above, also named *pedigree inbreeding* by Keller and Waller (2002), will be used in this chapter.

One of the mechanisms behind inbreeding depression can be described as follows: as a result of mutation, individuals can carry unique deleterious recessive alleles in the heterozygous state. This may not affect their overall fitness and, therefore, carriers of such alleles would be able to produce viable offspring. These recessive alleles can only be expressed when descendants of the same ancestor(s) mate with each other, leading to a homozygous state.

The fitness of offspring of consanguineous matings can consequently be compromised – varying from still–births, physical abnormalities to infertility and lower life expectancy (see Sect. 13.3). Since the first studies on inbreeding depression in zoo populations (Flesness 1977; Ralls and Ballou 1983; Ralls et al. 1979), many studies on this subject have followed (see e.g. Allendorf et al. 2013; Frankham et al. 2010).

Now, nearly 40 years after the first publications on inbreeding in zoo populations, it is generally accepted that inbreeding depression can jeopardise conservation of small populations, both in captivity and in the wild (see overviews in Allendorf et al. 2013; Frankham et al. 2010). Avoidance of or minimising inbreeding has been one of the main fundamental goals of the cooperative breeding programmes of international and regional zoo organisations, since their establishment in the early 1980s. Therefore, this chapter will focus on methods of detecting inbreeding depression in studbook populations.

13.2 Inbreeding Coefficients

The inbreeding coefficient (symbol: f) refers to the probability that a diploid individual inherits the same ancestral alleles at a given locus (Wright 1922). This probability can be computed from pedigree data, either by hand for small pedigrees or using computer software for larger populations. Figure 13.1 shows a pedigree of a full–sibling (brother and sister) mating. The ancestors, male 1 and female 2, are assumed to be unrelated and have genotypes *A1A2* and *A3A4*, respectively. The probability that individual 5 is *autozygous* for allele *A1* is estimated as follows:

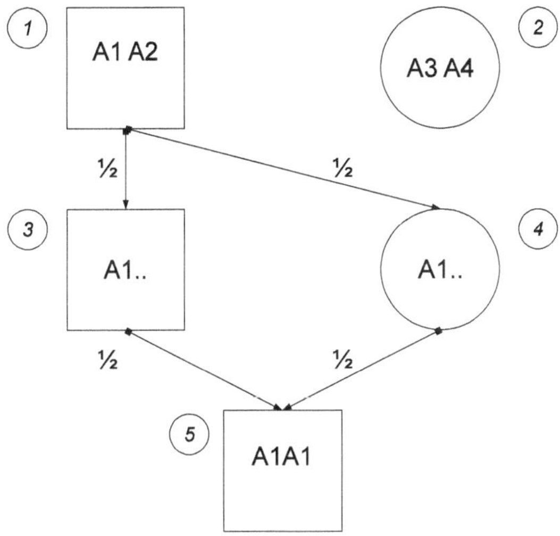

Fig. 13.1 Pedigree of full–sib mating. The small numbered *circles* are (hypothetical) IDs of individuals. *Squares* are males and (large) *circles* are females. The codes *A1* to *A4* refer to allelic variants at the same locus

1. The probability that either male 3 or female 4 inherits allele *A1* from ancestor 1 is $\frac{1}{2}$.
2. The probability that both male 3 and female 4 inherit allele *A1* is $\frac{1}{2} \times \frac{1}{2} = \frac{1}{4}$.
3. The probability that individual 5 inherits allele *A1* from one of its parents is $\frac{1}{2} \times \frac{1}{2} = \frac{1}{4}$.
4. The probability that individual 5 is homozygous for allele *A1* is $\frac{1}{4} \times \frac{1}{4} = \frac{1}{16}$.

The inbreeding coefficient (f) is the total probability that an individual is homozygous for one of the four ancestral alleles at a locus. For individual 5 in this example, it is $4 \times \frac{1}{16} = \frac{1}{4}$ or 0.25 (see Wright 1922). The inbreeding coefficient increases when inbreeding in subsequent generations continues.

Calculation of inbreeding coefficients, as illustrated in the previous example, can be a tedious and time–consuming job for complex pedigrees. Various techniques like *path analysis or chain-counting technique* and the *additive relationship matrix method* have been developed to calculate inbreeding coefficients of individuals in complex pedigrees (see Allendorf et al. 2013; Ballou 1983; Frankham et al. 2010). The additive relationship matrix method provides a straightforward method to calculate inbreeding coefficients regardless of the complexity of the pedigree.

However, whenever pedigree data are computerised (as in zoo managed studbooks), it is much more efficient to calculate inbreeding coefficients of individuals using software. For example, the *Quaas–Henderson computer algorithms* are based on the additive relationship matrix method (Henderson 1976; Quaas 1976, 1989). Meanwhile faster computer algorithms, especially those to manage large livestock pedigrees, have been designed (Colleau 2002; Sargolzaei et al. 2005).

It is important to realise that inbreeding coefficients are *probabilities*. Offspring of full–sib matings, in this example with four different ancestral alleles, actually have a probability of 0.75 to be heterozygous at this locus. However, inbreeding depression is not just a matter of expression of deleterious recessive alleles at a single locus. Genomes of higher organisms contain somewhere between 4,000 and 50,000 structural loci (Nei 1987). This means that the genome of an ancestor may contain recessive deleterious alleles at several loci, and, therefore, the probability of inbreeding effects increases. Sensitivity to inbreeding depression may vary with species or even with populations (Shields 1987). While the potential effects of inbreeding depression are known, it is difficult to predict which species or which population will be affected. Computerised studbook data offer the possibility of analysing potential inbreeding depression in the (studbook) population.

13.3 Measuring Inbreeding Depression

Inbreeding depression affects the fitness of individuals. Since fitness refers to the ability to produce viable offspring, it encompasses the whole life history of an individual, from as early as a zygote, embryo, to reproduction and successful rearing of offspring, until death. This means that inbreeding depression can act at different life stages. For example, an inbred embryo may not develop (and die); an inbred individual may not be fertile, an inbred female of a mammalian species may not be able to lactate and, therefore, will not be able to raise her offspring (*maternal effect*); or inbred individuals may have a lower life expectancy and, therefore, produce fewer offspring during their life–time than non-inbred individuals.

Unfortunately, detection of inbreeding depression in the prenatal stages may not be feasible, or at least not straightforward, in studbook populations. Studbooks generally only include individuals that were born or hatched (including premature births and stillbirths) (Thompson et al. 1997). This means that lethal inbreeding effects in the prenatal stages (including the period of incubation in egg–laying species) cannot be detected directly at the level of inbred individuals. Instead, indirect methods, such as assessing fecundity in the parent(s), are used to reveal such inbreeding effects.

Inbreeding depression in the prenatal stages can be detected in egg–laying (oviparous) species whenever data on clutch size, number of fertile eggs and hatching success are recorded in studbooks. It is important to study data from artificial incubation and incubation by the parents, separately. Artificial incubation introduces controlled environmental conditions (temperature and humidity) that may differ from parental incubation conditions.

The demographic tools that are available to analyse studbooks provide a wide range of options to detect inbreeding depression during the different life stages. These options include comparing infant mortality, survivorship, fecundity and overall fitness. Life table analyses provide data on age specific fecundity and mortality rates (Chap. 7). The distribution of these rates can be compared between inbred and non–inbred groups. Inbreeding depression does not necessarily act at a specific stage in life, such as the neonate or juvenile stages. The effects may be subtle, such as a slightly reduced fecundity in combination with a shorter life expectancy in inbred individuals. These effects could potentially be detected by studying reproductive values (see Chap. 7).

Application of statistical methods to detect inbreeding depression depends on the sample sizes of inbred and non–inbred groups. For example, the study of Ralls and Ballou (1983) on effects of inbreeding on infant mortality in species held at the National Zoo Washington, had to group all inbred individuals in one category, due to the small sample sizes at this single zoo collection. Analysing mortality in neonates or juveniles grouped in a single inbred category, provides the best method of detecting inbreeding depression in smaller studbooks. The sample sizes of these age (stage) groups are generally much larger than in older age groups (see Chap. 7). Sample sizes that are required for life table analyses have been discussed in Chap. 8.

Larger studbooks – read sample sizes – allow the inclusion of different levels of inbreeding in analyses. A study on the international studbook of Goeldi's monkeys (*Callimico goeldii*) compared infant survival with inbreeding coefficients. The study was based on 790 captive–born animals of which 111 were inbred (Lacy et al. 1993). The method to estimate lethal equivalents (Templeton and Read 1983) has a similar approach and can easily be applied to the same data set. These types of analyses are useful for deciding which levels of inbreeding – within the context that inbreeding cannot be completely avoided in small populations – can be tolerated. For example, inbreeding depression, measured as 30–day survival, in female Goeldi's monkeys was not only significant in offspring of full-sib matings ($f = 0.25$), but even in matings between first cousins ($f = 0.0625$) and pairings of less closely related individuals (Lacy et al. 1993).

13.4 Neonatal and Juvenile Mortality

Comparing infant mortality between inbred and non–inbred groups has proven to be an adequate method to detect inbreeding depression, even in relatively small data sets such as individual zoo collections (Ralls and Ballou 1983; Ralls et al. 1979). The definition of the "infant stage" depends on the life history of the species concerned. Ralls and Ballou (1983) used 180 days for ungulates and primates, and half the age of sexual maturity in small mammals. Results from analyses on lifespan and reproduction as presented in Chaps. 3 and 7 can be used to determine infant and other life stages.

First month (30 days) survival is often used in studies of inbreeding depression in zoo populations (e.g. Cassinello 2005; Lacy et al. 1993; Templeton and Read 1983).[2] The biological significance of this criterion largely depends on the natural history of a species (such as longevity and reproductive life–span). For example,

[2] Annual zoo inventories include 30 day survival of the newborns for each species. This "tradition" could be the basis for using this survival criterion.

studies on inbreeding depression in deer mice (*Peromyscus*) used survival rates until weaning at 20 days (Brewer et al. 1990).

A measure of inbreeding depression (δ) is the proportionate decline in fitness due to a certain amount of inbreeding (Frankham 2010):

$$\delta = 1 - \frac{\textit{fitness of inbred group}}{\textit{fitness of non} - \textit{inbred group}} \tag{13.1}$$

where fitness in general refers to neonatal or juvenile survival rates as discussed in the previous section.

13.4.1 Two–Way Contingency Table

The validity of δ is determined by the sample sizes of both groups. Differences in fitness (survival) between non-inbred and inbred groups can be tested by constructing *two–way contingency tables* and, depending on the sample size, applying the Fisher's exact test or Pearson's chi–square test (Siegel and Castellan 1988). The first comparative studies on inbreeding depression in zoo populations used the chi–square statistic (Ralls and Ballou 1983; Ralls et al. 1979).

Table 13.1 presents a contingency table for 6 months (juvenile) survival in inbred and non–inbred Nepalese red pandas born between the years 1977 and 2012. Both inbred and non–inbred parents are included in this analysis. The mean inbreeding coefficient ($\bar{F}$) in the inbred group is 0.047, which is between the level of *first cousin once removed* and *first cousin* matings.

The sample sizes are sufficiently large to apply the Pearson's chi–square test (Siegel and Castellan 1988). The null hypothesis is that 6 month survival does not differ between inbred and non–inbred groups. This test rejects the null hypothesis ($\chi^2 = 7.1, p = 0.008$), which means that juvenile survival in inbred Nepalese red pandas is lower than in non–inbred ones. Inbreeding depression (δ) for juvenile survival in offspring of non–inbred parents is 0.096 (Table 13.1). This

Table 13.1 Two–way contingency table to test for differences in 6 month survival in non–inbred and inbred ($\bar{F} = 0.047$) Nepalese red pandas born from 1977 to 2012

Offspring group	Litters	Survived	Died	Total	P_{180}
Inbred	922	966	597	1,563	0.618
Non–inbred	305	369	171	540	0.683
Total	1,227	1,335	768	2,103	

P_{180} is the survival rate after 180 days
Inbreeding depression $\delta_{[\bar{F}=0.047]} = 1 - \frac{0.618}{0.683} = 0.096$
Pearson's chi–square test with Yates' continuity correction: $\chi^2 = 7.1$, $df = 1$, $p = 0.0077$
Donner's adjustment: $\chi^2_D = 5.2$, $df = 1$, $p = 0.022$

means that fitness has been reduced by 9.6 % in inbred offspring due to an average inbreeding coefficient of $\bar{F} = 0.047$.

13.4.2 *Litter Effect*

The calculation of juvenile survival of red pandas can be subject to the litter effect i.e. pseudoreplication (Hurlbert 1984). This means that results of the Pearson's chi–squared test may not be valid. Several adjusted chi–squared tests for clustered binary data (e.g. survival in litters) have been developed (Song and Ahn 2003). The *Donner's test* is an adjusted Mantel–Haenszel test (Donner 1989, 1993) and implemented in the R package *aods3* (Lesnoff and Lancelot 2013). This test is applied to analysis of the red panda data.

The clustered binary tests estimate the variance *within* litters. This means that the data need to be provided as survivors and deaths per litter. The number of litters in inbred and non–inbred groups are 922 and 305, respectively. The adjusted chi–square for Donner's test (χ^2_D) is 5.2 (Table 13.1). The null hypothesis that survival in both groups does not differ is rejected as $p_{[5.2]} = 0.022 < 0.05$. This is the same conclusion as drawn from the Pearson chi squared test ($\chi^2 = 7.1$, $p = 0.008$).

The lower χ^2_D in Donner's test compared to Pearson's χ^2 shows that the litter data in red pandas is subject to pseudoreplication. In this example, both tests result in the same conclusion at a significance level (α) of 0.05. However, the effect of pseudoreplication on hypothesis testing can be illustrated by lowering α to 0.01 (99 % confidence level). The null hypothesis will *not* be rejected by Donner's test but will be rejected by Pearson's chi–squared test. This shows the importance of using adjusted chi–square tests for species with litters to avoid wrongly rejecting the null hypothesis ("there are no differences").

13.4.3 *Inbred and Non–Inbred Parents*

Analysis of inbreeding depression in the previous section did not distinguish between inbred and non–inbred parents. This offspring–based analysis may only reveal a part of inbreeding depression, as ancestral inbreeding effects in parents can play a role in survival (and fecundity) of offspring too (Ballou 1997; Lacy et al. 1996). For example, inbreeding depression may affect lactation in inbred females, and, therefore affect survivorship and even fecundity in their offspring.

A persistent misunderstanding is that offspring of inbred parents are inbred, and offspring of non–inbred parents are not inbred. Inbreeding can only occur in offspring of parents that are related, regardless of whether the parents themselves are inbred or not. For example, individuals 1 and 2 in Fig. 13.1 are unrelated and, therefore their offspring 3 and 4 are non–inbred. However, offspring of non–inbred parents 3 and 4, which are siblings, will result in inbred offspring.

Table 13.2 Effects of inbreeding in parents on 6 months survival in inbred and non–inbred Nepalese red pandas born from 1977 to 2012

Non-inbred					Inbred				
Parents	Litters	Survived	Died	P_{180}	Survived	Died	P_{180}	δ	p_{adj}
None inbred	540	307	141	0.685	289	195	0.597	0.128	0.022
Dam inbred[a]	687	62	29	0.681	677	403	0.627	0.08	0.363

P_{180} is the survival rate after 180 days, δ inbreeding depression, p_{adj} probability of adjusted χ^2 in Donner's adjusted Mantel–Haenszel test
[a]Sire is non–inbred

Table 13.2 presents the results of two–way contingency tables and the Donner's test for non–inbred parents and inbred dams with non-inbred sires. The difference in 6 months survival rate (P_{180}) between inbred and non–inbred offspring of non–inbred parents is significant ($p_{adj} = 0.022$).

The inbreeding depression in inbred offspring of non–inbred, but related, parents is $\delta = 0.128$ (Table 13.2). This is higher than $\delta = 0.096$ in inbred offspring of the parental group in which both inbred and non–inbred parents are combined (Table 13.1).

The Donner's test on offspring from inbred dams (and non–inbred sires) does not reject the hypothesis that survival rates in inbred and non–inbred offspring are the same. However, *not rejecting* is different from *accepting* a null hypothesis. The sample size of non–inbred offspring ($N = 62$) is only 8.4 % of inbred offspring ($N = 677$), which may be too small for rejecting the null hypothesis.

The above does not fully explain the results. If inbreeding has negative effects on females raising offspring then the survival rate in non–inbred offspring of inbred dams is expected to be lower than in non–inbred offspring of non–inbred parents. However, the differences between these groups are marginal in the red panda. Moreover, the survival rate in inbred offspring of inbred dams (0.627) is actually higher than inbred offspring of non–inbred parents (0.597) (see Table 13.2).

Various factors can play a role in the observed survival rates of inbred offspring of inbred dams. Since this is a chapter on inbreeding, it is tempting to look in the direction of (unintended) *purging*, i.e. deleterious recessive alleles are eliminated from the population through inbreeding, which increases homozygosity and thus exposure to selection (Frankham et al. 2010). Elimination of, particularly, lethal recessive alleles through purging is expected to improve survival in the next generation. This is reflected in survival of inbred offspring of inbred dams. But that explanation would be too simple.

The red panda studbook used in the analyses comprises 35 years of data, during which time husbandry practices improved. However, the Nepalese red panda pedigree is complex with inbreeding in the early years, and genetic management since 1977 (Princée 1988). This means that inbred offspring of non–inbred parents might refer to the early years, whereas non–inbred offspring of inbred parents to more recent periods.

Table 13.3 Effects of age groups and survival in inbred and non–inbred Nepalese red panda born to non–inbred parents from 1977 to 2012

Non-inbred					Inbred				
Age group (days)	Litters	Survived	Died	P_x	Survived	Died	P_x	δ	p^a_{adj}
0–30	540	340	108	0.759	346	138	0.715	0.054	0.216
1–30	520	340	76	0.817	346	119	0.741	0.085	0.035
1–7	520	363	53	0.873	400	65	0.860	0.010	0.656

P_x survival rate, p_{adj} probability of χ^2_D from Donner's adjusted Mantel–Haenszel test

13.4.4 Age Groups

Inbreeding depression may play a role at different stages in life and/or its effect can be overshadowed by other factors. Table 13.3 presents results of adjusted Mantel-Haenszel tests between inbred and non–inbred offspring of non–inbred parents for different age groups (in days).

Survival rates do not differ between inbred and non–inbred offspring when the survival age is changed from 180 to 30 days ($p_{adj} = 0.216$). However, these rates are statistically different ($p_{adj} = 0.035$) when deaths at date of birth (which includes stillbirths) are excluded from the sample (ages 1–30 days). Survival in the first week (days 1–7) does not show significant differences ($p_{adj} = 0.656$) between inbred and non–inbred offspring. This does not necessarily mean that inbreeding has no effect. Other factors, such as maternal care and husbandry practices, may prevail over inbreeding effects in the first week.

13.5 Lifespan

The previous section illustrated that inbreeding depression either has no effect in the early life of red pandas or can not be detected. Survival curves can provide more detail on inbreeding depression during different life stages than the two-way contingency tables.

The Kaplan–Meier product limit estimator (Kaplan and Meier 1958) and the Cox proportional hazards regression (Cox 1972) can be applied to test for differences between inbred and non–inbred groups. The Cox's model supports cluster (marginal) and frailty models, and is therefore preferred over the Kaplan–Meier estimator and logrank tests when testing for inbreeding effects in species with litters. Furthermore, this method can analyse both categorical data ("inbred" and "non-inbred") and numerical data. The latter means that the effect of increase in inbreeding coefficient on survival can be quantified. See Chap. 10 for more details on both methods.

Figure 13.2a presents Kaplan–Meier survival curves of inbred and non–inbred Nepalese red pandas born to non–inbred parents between 1977 and 2012. Stillbirths

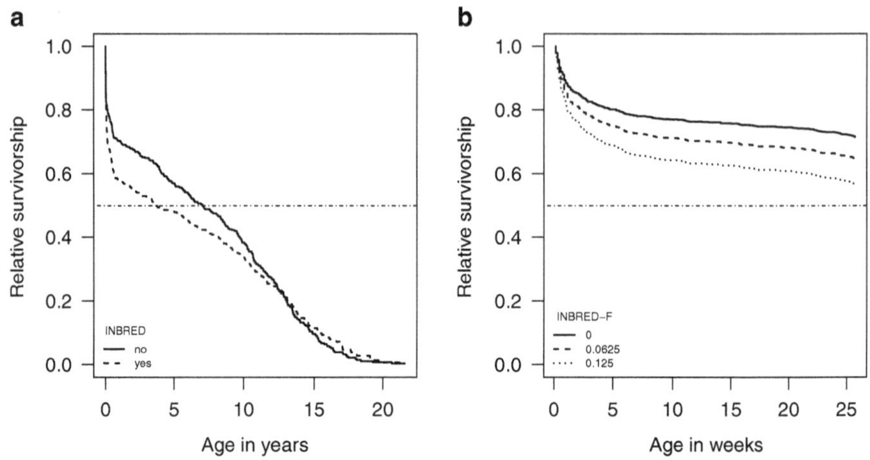

Fig. 13.2 (**a**) Kaplan–Meier survival curves of inbred and non–inbred Nepalese red pandas born to non–inbred parents between 1977 and 2012. Scale is years. (**b**) Predicted first year survival curves from Cox regression for inbreeding coefficients of 0, 0.0625 and 0.125 in the age group 1–180 days. The marginal (cluster) model is applied. Scale is weeks

and deaths at day of birth are excluded in this and further analyses in this section. Mortality in inbred red pandas during the early (juvenile) life stage is higher than in non–inbred red pandas. Survivorship in inbred individuals is slightly higher than in non–inbred individuals in the older ages (above 12 years).

The litter effect in red pandas does not allow the application of the logrank test on Kaplan–Meier survival curves (see Chap. 10). Therefore, Cox's model is preferred. Since major differences in survival between inbred and non–inbred groups occur in the juvenile period, it makes sense to continue with the 6 month period in analyses. As the results of two–way contingency tables show that inbreeding has an effect on survival (Table 13.2), the Cox regression model is used to predict the effects of different inbreeding coefficients.

Figure 13.2b presents survival curves for different inbreeding coefficients in the age group 1–180 days ($\approx$25 weeks). The predicted survival of red pandas is lower in more inbred offspring. The marginal model in which litters are considered as clusters is applied (see Chap. 10). These curves are predictions that are based on the estimated hazard ratio (*HR*) with the Cox regression model. The hazard ratio in this model is 67.3 ($p < 0.001$). This result can be accepted as the proportional hazard assumption is not violated i.e. $p = 0.103$ (see Chap. 10 for details on Cox regression).

The hazard ratio presented in the results of statistical software refers to a "one–unit" increase, which equals the maximum inbreeding coefficient of 1.0. The hazard ratios and particularly survival rates for the observed range of inbreeding coefficients are of more interest. These values can be estimated with Eqs. 10.3 and 10.5 (see Chap. 10). It is obviously more convenient to have statistical software which estimates (predicts) survival rates. For example, the survival rates for non–inbred and $F = 0.125$ groups at the end of the 180 day period are 0.714 and 0.565, respectively. This means that survival in non–inbred Nepalese red pandas is some 26 % higher than in the inbred group.

Cox's model has been used to test for inbreeding depression (e.g. Fredrickson et al. 2007; Ross–Gillespie et al. 2007; Sherwin et al. 2000). However, this method does not seem to be in common use and clearly deserves more attention, as it can not only handle pseudoreplication, but also provides detailed information on the effects of increase in inbreeding as well.

13.6 Lethal Equivalents

Lethal effects of inbreeding are not necessarily caused by expression of lethal recessive alleles at a single gene. They can be the result of a combination of detrimental alleles, named *lethal equivalents*:

> A lethal equivalent is a group of mutant genes of such number that, if dispersed in different individuals, they would cause on the average one death, e.g., one lethal mutant, or two mutants each with 50 per cent probability of causing death, etc. (Morton et al. 1956).

Lethal equivalents are estimated from survivorship to a given age, e.g. first month or first year, in groups of individuals with different levels of inbreeding (including non–inbred). Only individuals with parents that are both non–inbred are included in the analysis in order to exclude potential effects of inbreeding depression that act at the level of reproduction and parental care. The relation between survivorship and inbreeding is estimated according:

$$S_i = e^{-(A+Bf_i)} \tag{13.2}$$

where S_i is the survivorship of individuals with inbreeding coefficient of f_i, to a specified age, A is the survivorship of non-inbred individuals, and B is the rate of decline of survivorship with inbreeding (Morton et al. 1956).

Parameter A is often denoted as S_0. Parameter B is approximately equal to the number of lethal equivalents in a haploid genome (gamete) (Morton et al. 1956). The lethal equivalents in a diploid genome (zygote) are $2B$.

Equation 13.2 can be rewritten into a logarithmic format (Morton et al. 1956):

$$-\ln S_i = \ln A + Bf_i \tag{13.3}$$

This equation has the similar form of linear regression as described for exponential growth (see Eq. 6.2). This means that linear regression can be applied to estimate A and B (the slope). Since survivorship is a rate, its significance depends on the number of births (sample size) in each inbreeding category. Therefore, *weighted least–square regression* is applied (Morton et al. 1956; Templeton and Read 1983).

Table 13.4 Lethal equivalents (2*B*) in 30–day survival of Nepalese red pandas born to non–inbred parents from 1977 to 2012 using different methods. Confidence limits are between brackets. LM: weighted log–linear regression, MLE: maximum likelihood, GLM: generalised linear model, Robust: generalised linear model with robust (sandwich) variance estimator. See text for details

Method	Group	S_0	2*B*	*p*	Dispersion
LM[a]	F groups	0.80 [1.15, 1.37]	2.92 [−0.22, 6.08]	0.0675[b]	
MLE	F groups	0.81 [0.78, 0.84]	2.78 [0.94, 4.86]		
GLM	F groups	0.81 [0.77, 0.83]	3.02 [0.68, 4.06]	0.004[c]	2.35
Robust	Litters	0.81 [0.77, 0.84]	3.10 [1.0, 4.12]	0.003[c]	0.95

S_0 survival in non–inbred, 2*B* lethal equivalents, *p* probability of Students' or Wald's statistic
[a]Adjustment restricted to inbreeding categories with no survivors
[b]Students' test on regression coefficient *B*
[c]Wald's test on regression coefficient *B*

Groups with no survivors cause a problem in log transformation. Templeton and Read (1983) "bypassed" this problem by applying a small adjustment by increasing the number of survivors N_i with 1 and the number of births B_i with 2, for each inbreeding category (i):

$$S_i = \frac{N_i + 1}{B_i + 2} \tag{13.4}$$

This adjustment to the logarithmic model has been applied by other authors (e.g. Ralls et al. 1988). This method raised extensive discussion because of the bias that it creates in the estimated lethal equivalents (Kalinowski and Hedrick 1998; Lacy 1997; Templeton and Read 1998; Willis and Wiese 1997). However, this bias can be small, especially when the adjustment is restricted to inbreeding categories without survivors (Kalinowski and Hedrick 1998). The log–linear regression is weighted for the number of births per inbreeding category.

The results of the linear regression (LM) on 30–day survival in Nepalese red pandas are presented in Table 13.4. The number of lethal equivalents (2*B*) is 2.92. This linear model is not a good fit ($p > 0.05$), as indicated by the large confidence interval, and needs to be rejected.

Non–linear *maximum likelihood estimation (MLE)* to fit data to the untransformed model (Eq.13.2) has been recommended instead of linear regression (Kalinowski and Hedrick 1998). The data are not weighted. Figure 13.3 presents 30–day survivorship and level of inbreeding in Nepalese red pandas as estimated using MLE. The dashed line refers to the maximum likelihood fit. The results are presented in Table 13.4. The estimated number of lethal equivalents is 2.78. The confidence interval of 0.94–4.86 indicates that the MLE model is also far from a perfect fit.

Linear regression assumes that response variables are distributed normally, while survival/mortality is distributed binomially (see Chap. 7). Generalised linear models (*GLM*) support, among others, the binomial distribution and are applied to estimate lethal equivalents (e.g. Armstrong and Cassey 2007; Brekke et al. 2010; Jamieson et al. 2007).

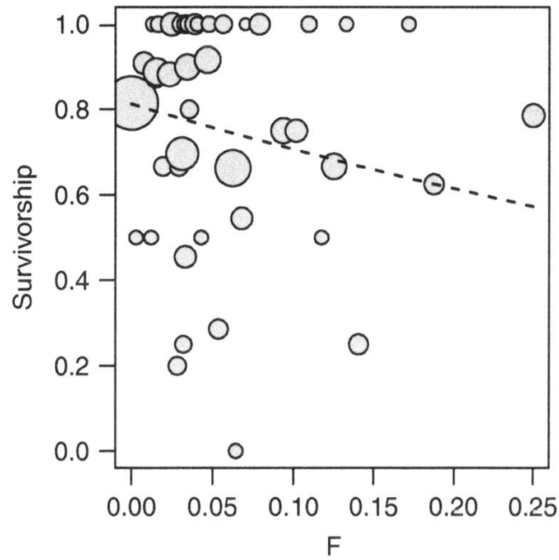

Fig. 13.3 Thirty day survival rates and levels of inbreeding (F) in Nepalese red pandas (1977–2012). Deaths at date of birth are excluded. The size of the *circles* is proportional to the number of births. The *dashed line* refers to the maximum likelihood fit ($B = 1.39$, $S_0 = 0.81$)

The survival data are presented in numbers of survivors and deaths per inbreeding category. The GLM model implemented in R (R Core Team 2015) carries out a weighted regression with sample sizes as weights (Crawley 2007). Table 13.4 presents the results of a binomial (GLM) regression on the same red panda data set as used in the LM and MLE analyses. The estimated values for S_0 and $2B$ are similar to those estimated by MLE methods, and coefficient B is significant ($p = 0.004$).

The significance level of the regression coefficient does not automatically mean that the GLM model should be "accepted". The *dispersion parameter* is 2.35, which means that the variation is larger than expected from the binomial model. This *overdispersion* can be due to pseudoreplication caused by, for example, litter or maternal effects.

Robust variance estimators, similar to the Donner' test and the marginal model (see Sects. 13.4.2 and 13.5), have been developed for generalised regression analyses.

The R package *robust* (Wang et al. 2014) is used in this section for estimating robust variance in a binomial regression model. The survival data are presented as survivors and deaths per litter. The results of the "Robust GLM" are in line with the MLE and GLM/F group model (Table 13.4). The Wald test ($p = 0.003$) and the dispersion parameter of 0.95 give this model preference over the others. The number of lethal equivalents associated to 30–day survivorship in the Nepalese red panda studbook population is $2B = 3.10$, following the robust model.

The GLM (binomial) regression is not restricted to grouped data, but can handle individual data as well (e.g. Brekke et al. 2010). The survival data are presented as binary data (yes/no), i.e. an individual died or survived. This means that details that refer to individuals can be included as variables. The *generalised linear mixed models (GLMM)* is an extension to the GLM model which supports random factors, such as litter or location of birth, to be included in the regression model (e.g. Hoeck et al. 2015).

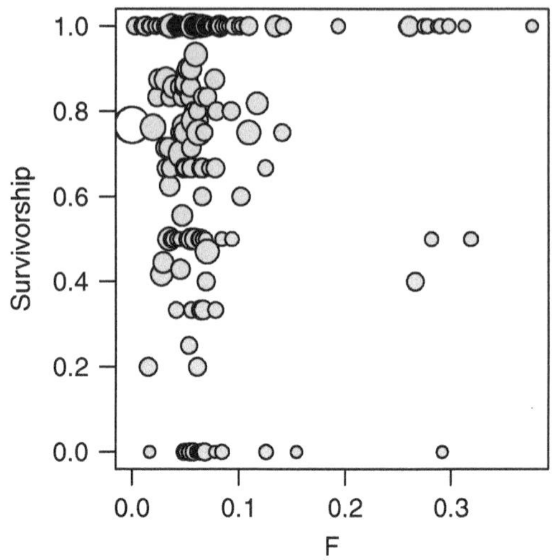

Fig. 13.4 Thirty day survival rates of offspring born to inbred female Nepalese red pandas. Deaths at date of birth are excluded. The size of the *circles* is proportional to the number of births

13.7 Maternal Inbreeding

The analyses that have been presented in the previous section focus on offspring that are inbred. This may only reveal a part of inbreeding depression, as ancestral inbreeding effects can play a role in survival and fecundity too (Ballou 1997; Lacy et al. 1996). For example, inbreeding depression may affect lactation in inbred females, and, therefore affect fitness of their offspring. However, the opposite effect occurs too as purging can reduce inbreeding depression affecting offspring (Ballou 1997).

Figure 13.4 presents 30–day survival rates of offspring born to inbred female Nepalese red pandas (offspring that died at date of birth are excluded). Robust regression analyses did not reject the null hypothesis i.e. no inbreeding depression in females that affected neonatal survival could be detected.

13.8 Fecundity

Inbreeding depression is not restricted to mortality, but can affect reproductive success as well. For example, *cryptorchidism* (undescended testes) has been observed in inbred Florida panthers (*Puma concolor coryi*) (Mansfield and Land 2002) and in inbred Scandinavian grey wolves (*Canis lupus*) (Räikkönen et al. 2013). Poor sperm quality due to inbreeding has been observed in the Mexican grey wolf (*Canis lupus baileyi*) (Asa et al. 2007). A comparative study of 20 outbred (non–inbred) populations of endangered and non–endangered species showed a negative relationship between sperm quality and decreasing levels of heterozygosity (Fitzpatrick and

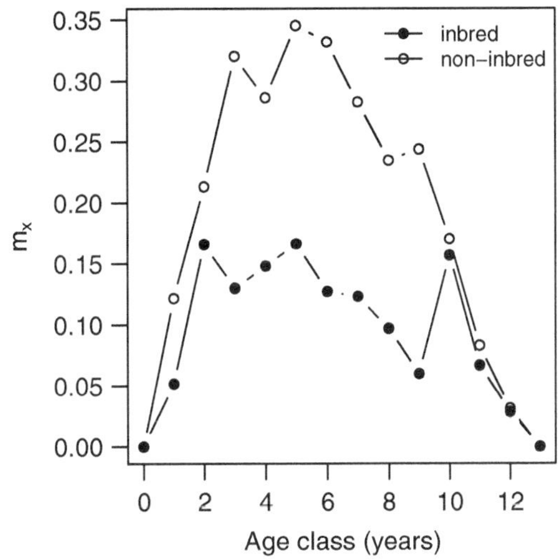

Fig. 13.5 Fecundity rates (m_x) in inbred (•) and non–inbred (∘) female Nepalese red pandas (data from 1977 to 2012)

Evans 2009). Reduced litter size was observed in inbreeding experiments on white–footed mice (*Peromyscus leucopus*) (Brewer et al. 1990; Lacy et al. 1996).

Age–structured fecundity tables (Chap. 7) can be used to analyse fecundity in inbred and non–inbred adults. The logrank test (Eq. 7.9) is used to test the (null) hypothesis that both groups are the same.

Figure 13.5 presents fecundity rates in inbred and non–inbred female Nepalese red pandas. Fecundity rates in adult non–inbred females are almost twice those in inbred females. These differences are significant, i.e. the null hypothesis is rejected ($\chi^2 = 93.5, p = 0, df = 1$) However, one should be cautious with the interpretation of results, as reproduction of animals in captivity is managed by humans. See Sect. 13.10 for potential data artefacts.

13.9 Reproductive Value

Age specific reproductive values (V_x) predict the number of offspring that an individual at a given age is expected to produce until the end of its life (see Sect. 9.3). They can be considered as a measure of fitness (e.g. Caswell 2001). Reproductive values have been used to detect inbreeding depression in the Nepalese red panda (Glatston and Princée 1993).

Absolute reproductive values are preferred as they refer to expected numbers of offspring. Furthermore, the value for age class 0 equals the net reproductive growth (R_0). Figure 13.6 shows absolute reproductive values for inbred and non–inbred female Nepalese red pandas estimated from class–based life tables (data 1977–2012).

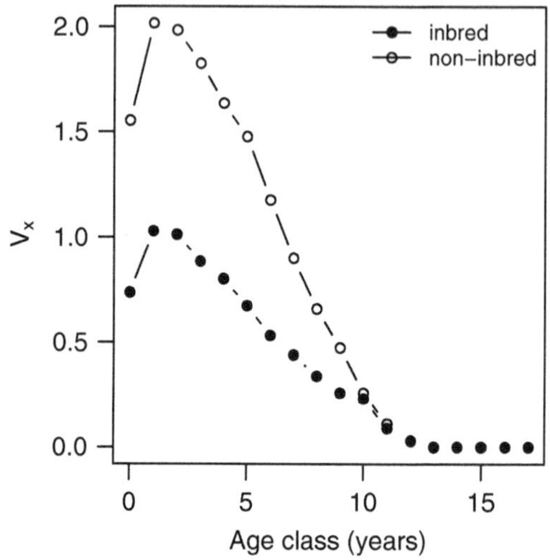

Fig. 13.6 Absolute reproductive value in inbred (•) and non–inbred (∘) female Nepalese red pandas (data from 1977 to 2012)

Non–inbred females have a higher fitness (roughly twice) than inbred females until both reach the post reproductive age. The effect of inbreeding can be considered dramatic when the net reproductive rates are taken into consideration: the non–inbred group has the potential to increase by 55 % per generation time ($R_0 = 1.55$), whereas the inbred group would decline by 26 % ($R_0 = 0.74$). The generation time for both groups is around 5 years and 3 months. However, the same caution in interpretation of results as mentioned for fecundity rates applies to differences in reproductive values between groups.

13.10 Data Artefacts

Life expectancy of individuals in zoos can increase over time as a result of improved husbandry (see Sect. 3.2). Developing husbandry guidelines is a fundamental part of cooperative breeding programmes. It may be anticipated that life expectancy would increase after such a programme started. This may lead to a situation in which juvenile mortality in the group of non–inbred individuals, but which were born in the initial years of the programme's establishment, is higher than in inbred individuals born later. However, the opposite situation can occur too, as early data may only include individuals which survived and which were relevant for the current population. This means that the period of time included in analyses of viability should be carefully selected. The date (year) that the first inbred individual was born could be used as the start of the "time window".

Improved husbandry is expected to have the same impact on reproductive success as discussed above for life expectancy. However, fecundity rates are also affected by breeding management. Mating opportunities under captive conditions are managed by humans. This may have affected reproductive success of inbred animals as the risks associated with inbreeding were interpreted by some zoo staff as if inbred animals were "inferior". This is a misconception, as being homozygous for a deleterious recessive allele is a matter of probability, as explained in Sect. 13.2. Population management programmes, as well as population genetic courses for zoo staff, have clearly changed this attitude. However, this misconception is persistent and may still affect inbred individuals belonging to species that are not managed.

If inbred animals were excluded from reproduction then it is likely this artefact occurred at the onset of population management programmes in zoos in the 1980s, as a reaction to the first publications of inbreeding effects several years earlier (e.g. Flesness 1977). Therefore it is recommended to examine studbooks to see whether both inbred and non–inbred animals reproduced during the early 1980s.

Reduced reproduction is a management tool to maintain populations within the available space (carrying capacity). Clearly fecundity rates and reproductive values will decline and can not be used in tests for inbreeding depression.

It should be clear that occurrence of inbreeding depression in populations, especially of endangered species, can jeopardise their conservation. Thus the general management approach in cooperative breeding programmes is to avoid or minimise inbreeding in captive populations (Ballou and Lacy 1995; Princée 1995). This can actually lead to a situation where studbooks do not have sufficient data on inbred individuals to test for inbreeding depression (Kalinowski and Hedrick 1999).

References

Allendorf FW, Luikart G, Aitken SN (2013) Conservation and the genetics of populations, 2nd edn. Wiley–Blackwell, Chichester

Armstrong DP, Cassey P (2007) Estimating effect of inbreeding on survival. Anim Conserv 10:487–492

Asa C, Miller P, Agnew M, Rebolledo JAR, Lindsey SL, Callahan M, Bauman K (2007) Relationship of inbreeding with sperm quality and reproductive success in Mexican gray wolves. Anim Conserv 10(3):326–331

Ballou JD (1983) Calculating inbreeding coefficients from pedigrees. In: Schonewald-Cox et al. (1983), pp 509–520

Ballou JD (1997) Ancestral inbreeding only minimally affects inbreeding depression in mammalian populations. J Heredity 88:169–178

Ballou JD, Lacy RC (1995) Identifying genetically important individuals for management of genetic variation in pedigreed populations. In: Ballou et al. (1995), pp 76–111

Ballou JD, Gilpin M, Foose TJ (eds) (1995) Population management for survival and recovery: analytical methods and strategies in small population conservation. Methods and cases in conservation science series. Columbia University Press, New York

Brekke P, Bennet PM, Wang J, Pettorelli N, Ewen JG (2010) Sensitive males: inbreeding depression in an endangered bird. Proc R Soc B 277:3677–3684

Brewer BA, Lacy RC, Foster ML, Alaks G (1990) Inbreeding depression in insular and central populations of *Peromyscus* mice. J Heredity 81:257–266

Cassinello J (2005) Inbreeding depression on reproductive performance and survival in captive gazelles of great conservation value. Biol Conserv 122:453–464

Caswell H (2001) Matrix populations models: construction, analysis, and interpretation, 2nd edn. Sinnauer Associates, Sunderland

Colleau JJ (2002) An indirect approach to the extensive calculation of relationship coefficients. Genet Sel Evol 34(4):409–421

Cox DR (1972) Regression models and life-tables. J R Stat Soc Ser B (Methodol) 34(2):187–220

Crawley MJ (2007) The R book. Wiley, Chichester

Darwin C (1876) The origin of species by means of natural selection or the preservation of favoured races in the struggle for life, 6th edn. John Murray, London

Donner A (1989) Statistical methods in ophthalmology: an adjusted chi-square approach. Biometrics 45(2):605–611

Donner A (1993) The comparision of proportions in the presence of litter effects. Prev Vet Med 18(1):17–26

Fitzpatrick JL, Evans JP (2009) Reduced heterozygosity impairs sperm quality in endangered mammals. Biol Lett 5(3):320–323

Flesness NR (1977) Gene pool conservation and computer analysis. Int Zoo Yearb 17:77–81

Frankham R (2010) Challenges and opportunities of genetic approaches to biological conservation. Biol Conserv 143(9):1919–1927

Frankham R, Ballou JD, Briscoe DA (2010) Introduction to conservation genetics, 2nd edn. Cambridge University Press, Cambridge

Fredrickson RJ, Siminski P, Woolf M, Hedrick PW (2007) Genetic rescue and inbreeding depression in Mexican wolves. Proc R Soc Lond B Biol Sci 274(1623):2365–2371

Glatston AR, Princée FPG (1993) A global masterplan for the breeding of *Ailurus fulgens* in captivity. Technical report, Stichting Koninklijke Rotterdamse Diergaarde (Rotterdam Zoo), Rotterdam

Henderson CR (1976) A simple method for computing the inverse of a numerator relationship matrix used in prediction of breeding values. Biometrics 32:69–83

Hoeck PE, Wolak ME, Switzer RA, Kuehler CM, Lieberman AA (2015) Effects of inbreeding and parental incubation on captive breeding success in Hawaiian crows. Biol Conserv 184:357–364

Hurlbert SH (1984) Pseudoreplication and the design of ecological field experiments. Ecol Monogr 54(2):187–211

Jamieson IG, Tracy LN, Fletcher D, Armstrong DP (2007) Moderate inbreeding depression in a reintroduced population of North Island robins. Anim Conserv 10(1):95–102

Kalinowski ST, Hedrick PW (1998) An improved method for estimating inbreeding depression in pedigrees. Zoo Biol 17:481–497

Kalinowski ST, Hedrick PW (1999) Detecting inbreeding depression is difficult in captive endangered species. Anim Conserv 2:131–136

Kaplan EL, Meier P (1958) Nonparametric estimation from incomplete observations. J Am Stat Assoc 53:457–481

Keller LF, Waller DM (2002) Inbreeding effects in wild populations. Trends Ecol Evol 17(5):230–241

Lacy RC (1997) Importance of genetic variation to the viability of mammalian populations. J Mammal 78:320–335

Lacy RC, Petric A, Warneke M (1993) Inbreeding and outbreeding depression in captive populations of wild species. In: Thornhill N (ed) The natural history of inbreeding and outbreeding. University of Chicago Press, Chicago, pp 352–374

Lacy RC, Alaks G, Walsh A (1996) Hierarchical analyses of inbreeding depresssion in *Peromyscus polionotus*. Evolution 50(6):2187–2200

Lesnoff M, Lancelot R (2013) aods3: analysis of overdispersed data using S3 methods. Aods3 package version 0.4-1

Mansfield KG, Land ED (2002) Cryptorchidism in Florida panthers: prevalence, features, and influence of genetic restoration. J Wildl Dis 38(4):693–698

Morton NE, Crow JF, Muller HJ (1956) An estimate of the mutational damage in man from data on consanguineous marriages. Proc Natl Acad Sci USA 42(11):855–863

Nei M (1987) Molecular evolutionary genetics. Columbia University Press, New York

Princée FPG (1988) Genetic variation in the zoo population of the red panda subspecies *Ailurus fulgens fulgens*. Zoo Biol 7(3):219–231
Princée FPG (1995) Overcoming the constraints of social structure and incomplete pedigree data through low-intensity genetic management. In: Ballou et al. (1995), pp 124–154
Quaas RL (1976) Computing the diagonal elements and inverse of a large numerator relationship matrix. Biometrics 32:949–953
Quaas RL (1989) Transformed mixed model equations: a recursive algorithm to eliminate A^{-1}. J Dairy Sci 72(7):1938–1941
Räikkönen J, Vucetich JA, Vucetich LM, Peterson RO, Nelson MP (2013) What the inbred Scandinavian wolf population tells us about the nature of conservation. PLoS ONE 8(6):e67218
Ralls K, Ballou J (1983) Extinction: lessons from zoos. In: Schonewald-Cox et al. (1983), pp 164–184
Ralls K, Brugger K, Ballou J (1979) Inbreeding and juvenile mortality in small populations of ungulates. Science 206:1101–1103
Ralls K, Ballou JD, Templeton A (1988) Estimates of lethal equivalents and the cost of inbreeding in mammals. Conserv Biol 2(2):185–193
R Core Team (2015) R: a language and environment for statistical computing. R Foundation for Statistical Computing, Vienna. https://www.R-project.org/
Ross–Gillespie A, O'Riain MJ, Keller LF (2007) Viral epizootic reveals inbreeding depression in a habitually inbreeding mammal. Evolution 61(9):2268–2273
Sargolzaei M, Iwaisaki H, Colleau JJ (2005) A fast algorithm for computing inbreeding coefficients in large populations. J Anim Breed Genet 122:325–331
Schonewald-Cox CM, Chambers SM, MacBryde B, Thomas L (eds) (1983) Genetics and conservation: a reference for managing wild animal and plant populations. Biological conservation. Benjamin/Cummings, Menlo Park
Sherwin WB, Timms P, Wilcken J, Houlden B (2000) Analysis and conservation implications of koala genetics. Conserv Biol 14(3):639–649
Shields WM (1987) Dispersal and mating systems: investigating their causal connections. In: Chepko-Sade BD, Halpin ZT (eds) Mammalian dispersal patterns: the effects of social structure on population genetics. University of Chicago Press, Chicago, pp 3–24
Siegel S, Castellan NJ Jr (1988) Nonparametric statistics for the behavioural sciences, 2nd edn. McGraw-Hill, New York
Song JX, Ahn CW (2003) An evaluation of methods for the stratified analysis of clustered binary data in community intervention trials. Stat Med 22(13):2205–2216
Templeton AR, Read B (1983) The elimination of inbreeding depression in a captive herd of Speke's gazelle. In: Schonewald-Cox et al. (1983), pp 241–261
Templeton AR, Read B (1998) Elimination of inbreeding depression from captive populations: Speke's gazelle: validity of the original statistical analysis and confirmation by permutation testing. Zoo Biol 17(2):77–94
Thompson SD, Earnhardt JM, Turner-Erfort G (1997) Guidelines for data entry and maintenance of North American regional studbooks. Lincoln Park Zoo, Chicago
Wang J, Zamar R, Marazzi A, Yohai V, Salibian-Barrera M, Maronna R, Zivot E, Rocke D, Martin D, Maechler M, Konis K (2014) Robust: robust library. R package version 0.4-16
Willis K, Wiese RJ (1997) Elimination of inbreeding depression from captive populations: Speke's gazelle revisited. Zoo Biol 16:9–16
Wright S (1922) Coefficients of inbreeding and relationship. Am Nat 56:330–338

Chapter 14
Genetic Drift and Simulations

Abstract Genetic drift is a random process that can lead to the fixation of alleles, but at the cost of losing alleles, especially those with low frequencies, and to increased homozygosity in the population. This process is described for an idealised population and its effect illustrated on different population sizes. The founder effect, a similar random process, is also described. The conditions of the idealised population are rarely met in real populations. Therefore, in order to estimate genetic drift it is necessary to make adjustments, for example by estimating effective population size (N_e). Methods to estimate N_e in populations with unequal sex–ratio in breeders, e.g. herd species, and with unequal family sizes, are provided. Problems in estimating genetic drift due to generation overlap are also discussed. Effective population size can also be estimated from pedigrees. This method is illustrated with the Nepalese red panda as example. Computer programs that simulate Mendelian segregation in pedigrees (gene dropping) estimate the expected genetic drift and are not hindered by unequal sex–ratio, unequal family size or generation overlap. The general methodology of gene dropping is described. Software implementations can differ in complexity, e.g. from single locus (two alleles) to multiple loci (multiple alleles) on chromosomes and crossing over. The effects of simple and complex models on genetic loss in the wolverine population were evaluated in the context of model selection. Effective population sizes are estimated in Nepalese red pandas from expected genetic drift between census intervals.

14.1 Introduction

Observed and expected heterozygosity are important measures for the genetic management of small populations (see Chap. 12). An increase in homozygosity is considered as a loss in genetic variation. Consanguineous matings increase the probability that an individual is *homozygous by descent*, and, consequently, results in an increase in homozygosity in the population.

Individuals can also be homozygous for alleles which do not originate from the same ancestor i.e. *homozygous by kind or state*. A random process named *genetic drift* can lead to fixation of alleles, and this also increases the probability that an individual is homozygous (for a non–ancestral allele).

F.P.G. Princée, *Exploring Studbooks for Wildlife Management and Conservation*,
Topics in Biodiversity and Conservation 17, DOI 10.1007/978-3-319-50032-4_14

It can be confusing that increase in homozygosity due to consanguineous matings and genetic drift are both named "inbreeding", but they are interpreted differently. The first type of inbreeding originates from the practice of livestock breeding in which it is considered important to asses the probability that an individual inherits recessive alleles (either deleterious or advantageous) from ancestors (Wright 1922).

Increase in homozygosity in a population can be due to genetic drift i.e. fixation of alleles. Inbreeding in this context is defined as matings between individuals who are more closely related than they would be if they had been chosen at random from a population (Crow and Kimura 1970). This means that inbreeding refers to the increase in homozygosity in the population relative to random mating (*Hardy–Weinberg*) proportions.

14.2 Genetic Drift and Founder Effect

Population genetic models often assume an *idealised* population. The main characteristics of this hypothetical population are (1) the number of breeding individuals is constant in each generation; (2) there is no generation overlap; (3) there is random mating; and (4) family size (number of offspring) follows a Poisson distribution (i.e. $mean = variance = 2$) (Frankham et al. 2010). This means that each individual in the ideal population has an equal chance of mating (twice) and passing its genes to the next generation. However, this also means that individuals have equal chances of *not* mating, or of producing only a single offspring, and thus none or only half of its genes are passed to the next generation. Allele frequencies in the offspring generation may therefore differ from the parental generation. This random sampling error of gametes per generation is called *genetic drift*.

The effect of genetic drift on genetic variation can be estimated by considering an ideal population of N individuals (of a diploid species) as a zygote pool of $2N$ gametes from which $2N$ gametes are sampled (with replacement). Gene diversity (symbol: H_e) at generation $t + 1$ is expected to be:

$$H_{e[t+1]} = (1 - \frac{1}{2N})H_{e[t]} \tag{14.1}$$

where N is the population size (Wright 1931).

Genetic drift is the proportional loss of $1/(2N)$ in gene diversity per generation. Gene diversity is also lost when establishing a new population, due to a similar sampling error to that causing genetic drift. A founder group of N individuals represents a fraction of $1 - 1/(2N)$ of the gene diversity in the original population. This process is named the *founder effect*.

Figure 14.1 shows the combined effect of population size on founder effect and genetic drift in small (ideal) populations. Values of gene diversity are presented as proportional to the original (wild) gene diversity. This figure illustrates the problems of maintaining small populations of endangered species. The smaller populations not only start with less gene diversity, but also lose more gene diversity per generation than large populations.

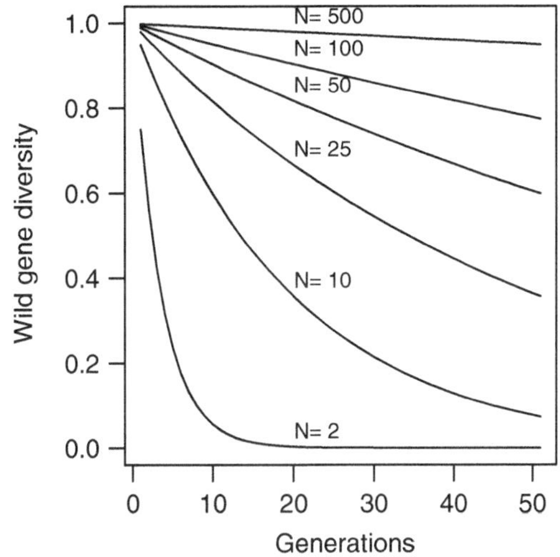

Fig. 14.1 Effect of population size (N) on founder effect and genetic drift in ideal populations

14.3 Effective Population Size

Real populations, especially of species with complex social systems, usually do *not* meet the assumptions of idealised populations. Unequal sex–ratio reduces the number of individuals (of the sex with the lowest number) that contributes genes to the next generation. This results in a genetic loss that is larger than predicted by genetic drift based on population size (Eq. 14.1).

Wright (1931) introduced the concept of *effective population size* (symbol: N_e) to estimate genetic drift in populations that deviate from the ideal population. This measure can be described as the number of individuals in an idealised population that would have the same genetic drift as an actual population.

14.3.1 Unequal Sex–Ratio

Sex–ratio in the breeding population is not necessarily equal, e.g. in species that live in harem groups. The effective population size (N_e) for unequal sex–ratio is calculated as:

$$N_e = \frac{4N_m N_f}{N_m + N_f} \tag{14.2}$$

where N_m and N_f, are the numbers of sexually mature males and females, respectively (Wright 1931).

14.3.2 Unequal Family Size

The model in Eq. 14.2 is based on the assumptions that family size follows a Poisson distribution, and that population size is constant. In reality, however, the variance in family size can differ; for example some individuals may never be in a breeding condition; or the population is growing i.e. mean family size is increasing. Furthermore, management to equalise reproductive success in captive populations can result in near zero variance in family size.

The *variance effective population size* (symbol: $N_{e(v)}$) (Kimura and Crow 1963) adjusts for both unequal sex–ratio and distribution of family size that deviates from a Poisson distribution. The variance effective size is calculated in two steps. First the effective number is calculated for the sexes separately (here illustrated for males N_{me}):

$$N_{me} = \frac{N_m \bar{k_m} - 1}{\bar{k_m} + (\sigma^2_{k_m}/\bar{k_m}) - 1} \tag{14.3}$$

where N_m is the number of males, $\bar{k_m}$ and $\sigma^2_{k_m}$ are the mean number and variance in offspring (both sexes) among sexually mature males (Kimura and Crow 1963).

The female variance effective population (N_{fe}) is calculated by substituting the appropriate female parameters in Eq. 14.3. The variance effective size is then calculated by substituting male and female numbers in Eq. 14.2 for the respective effective numbers.

14.3.3 Generation Overlap

Generation overlap is not taken into account in either of the models described above. Demographic models to adjust for generation overlap incorporate generation time (e.g. Engen et al. 2005; Felsenstein 1971; Lande and Barrowclough 1987). Practical application of these mathematical models to wild populations is subject to discussion, in favour of the use of computer simulation models in which population development and genetic change over time are combined (Allendorf et al. 2013; Frankham et al. 2010).

This discussion of generation overlap models also applies to captive populations, as generations are defined in relation to the founders (see Chap. 12). Therefore, generation overlap is not only caused by overlap in reproductive life of (grand)parents and offspring, but by imports of founders over longer periods of time as well. Section 14.5 presents genetic simulation models that can estimate genetic drift between captive–born generations and census intervals from pedigree data (e.g. Princée 1988, 1998). Effective population size can be estimated from simulation results following Eq.14.1.

14.4 Estimating N_e from Pedigrees

The number of offspring that has been produced by an individual over its lifetime can be extracted from pedigree data. This means that variance effective size can be calculated for any group of individuals, e.g. captive–born generation group, individuals that reproduced during one generation time interval, or the living population (e.g. Ballou et al. 2010; Princée 1988). These estimates provide an insight into the effects of demographic processes on genetic variation and thus into the feasibility of management measures to increase N_e/N ratios in order to minimise genetic loss.

A relatively simple method of estimating variance effective size of the living population is to estimate $\bar{k}$ from the number of offspring born to (living) males and females that have reproduced and to assume a Poisson distribution of family sizes (i.e. $\sigma_k^2 = \bar{k}$). This demographic method is implemented in the program *PMx* (R. Lacy, personal communication). The advantage of this method is that $N_e(v)$ can be estimated without detailed pedigree data, which otherwise would be necessary to estimate σ_k^2.

A further refinement to this method is to include non–breeders that have reached sexual maturity at census. Offspring that died before reaching sexual maturity are excluded from birth counts; and variance family size is estimated from pedigree data (see Ballou and Foose 1996; Lande and Barrowclough 1987). Table 14.1 presents variance effective size for the living population of Nepalese red pandas (31 Dec 2012) based on this refined method. Age at sexual maturity is ≈2 years. The values between brackets refer to breeders only i.e. $N = N_{breeder}$.

The effective population size of 151 results in a genetic drift of 0.33 % per generation (time). Whether this loss is acceptable depends on management goals with respect to the minimal amount of original (wild) genetic variation to retain, the time–frame and carrying capacity (zoo space). The total living population (N) is 402 individuals, resulting in an N_e/N ratio of 0.38.

Excluding non–breeders from the living population has only marginal effects on effective numbers per sex, and none on total N_{ev}. This means that identifying breeders (and their offspring) seems to be sufficient to estimate effective size.

Table 14.1 Variance effective population size of living sexually mature Nepalese red pandas on 31 December 2012. Age at sexual maturity is 2 years for both sexes. Values between brackets refer to variance effective size in living breeders

Sex	N	$N_{breeder}$	k	$\bar{k}$	σ_k^2	$N_{e(v)}$	N_e/N
Male	201	86	313	1.56 (3.64)	6.09 (6.66)	69.85 (69.82)	0.35 (0.81)
Female	201	91	326	1.62 (3.58)	5.42 (4.93)	82.04 (82.07)	0.41 (0.90)
Total	402	177				150.91 (150.91)	0.38 (0.84)

N number of sexually mature individuals, $N_{breeder}$ number of breeders, k total number of sexually mature offspring, $\bar{k}$ mean number of sexually mature offspring per individual, σ_k^2 variance in family size, $N_{e(v)}$ variance effective size, N_e/N (variance) effective size/population size ratio

However, including non–breeders provides a more realistic N_e/N ratio to determine the required population size and/or to evaluate whether reproductive success needs to be improved.

Although effective size can be adjusted for generation overlap, demographic methods are likely to result in overestimates due to effects of non–random mating and, in the case of captive populations, overlap in captive–born generations. Comparing estimated values with results of genetic simulation models is therefore recommended.

14.5 Genetic Simulation Models

Genetic drift in populations with unequal sex–ratio, non–random distribution of family size and generation overlap can be estimated with various mathematical models of effective population size. These models, however, assume that mating in these (real) populations is random. Furthermore, genetic drift estimated from N_e values can be inaccurate due to overlap in captive–born generations (see Sect. 14.2).

The random nature of Mendelian segregation makes it suitable for applying *Monte Carlo methods* to study processes such as genetic drift in pedigreed (studbook) populations. This technique is named *gene dropping* and does not have the limitations of mathematical models in estimating genetic drift as mentioned above.

The first computer programs to analyse genetic structure of pedigreed populations were already developed in the 1960s (e.g. Gilbert and Hammel 1966; MacCluer 1967). However, the introduction of personal computers in the 1980s provided a wider access to software, and initiated development of simulation models for studbook populations (e.g. MacCluer et al. 1986; Princée 1988).

The basic methodology of gene dropping can be described in the following steps:

1. Assign genotypes to founders, either from a pre-determined list or sample from allele distributions using Monte Carlo methods.
2. Assign genotypes to descendants of founders in the pedigree using Monte Carlo methods to simulate Mendelian segregation.
3. Compute measures of genetic variation from genotypes in groups of descendants e.g. generation groups and living population.
4. Iterate these steps e.g. in the order of 1,000–10,000 times.
5. Compute mean, variance, percentiles etc. over measures obtained per iteration (run).

The original gene dropping model ("gene drop") (MacCluer et al. 1986) assumes a single neutral locus and assigns two unique allelic variants to each founder. Since each founder has two unique alleles, the (observed) heterozygosity in the founder group is 1.0. Gene diversity in the founder population is estimated using Eq. 12.4.

The gene drop model is implemented in the pedigree analysis program *GENES* (Lacy 1993). More recent versions of this simulation program are embedded in the population analyses program PMx (Ballou et al. 2011). Both programs are generally associated with studbooks that have been created in SPARKS (Scobie and Flesness 1989). The gene drop model is therefore widely used in management of zoo populations.

Genome models implemented in gene dropping software can be more complex than a single locus. For example, the program *GeneFlow* (Princée 1988) was designed to handle multiple independent autosomal loci. Monte Carlo methods are used to draw random genotypes of founders from hypothetical and/or empirical allele frequencies in an infinitely large (wild) population, at the start of each iteration (run). This means that GeneFlow also simulates the stochastic nature of the founder effect.

The program *ChromoFlow* (Princée 1998) is the successor of GeneFlow. This program simulates transmission of multiple autosomal chromosomes on which various linked genes can be located. Numbers of loci, numbers of allelic variants and their frequencies can be manipulated as described for the GeneFlow model. Monte Carlo methods are used to sample homologues for descendants of founders based upon Mendelian segregation of the parental homologues, in each simulation run.

ChromoFlow accommodates recombination by crossing over during meiosis, following Haldane's model (Haldane 1919; Stam 1979). Therefore, the location of each locus in the genome is defined, i.e. chromosome number and map distance in centi–Morgans from either centromeric or telomeric end. Genetic crossing over is applied prior to Mendelian segregation of the parental homologues for each descendant and in each simulation run. Genome and genetic composition models can be hypothetical and/or based on empirical data.

14.5.1 Which Model?

The difference in complexity of gene dropping models triggers the question "Which model would be best to use?". Complex models are not necessarily better than simpler ones, depending on the goal (see e.g. Starfield and Bleloch 1991). The model ChromoFlow was designed by Princée (1998) to study the impact of assumptions regarding genome and genetic composition models. Results from that study showed that mean values of genetic variation computed over iterations are similar between models, but that variances can differ greatly, especially when rare alleles are involved.

Simulation experiments with the ChromoFlow program (Princée 1998) on the wolverine EEP population have been carried out to illustrate the effects of assumptions regarding allele frequencies in a "single locus" model. Mean and variance of gene diversity were estimated over 10,000 runs. The living population in 2000 represents descendants of eight founders.

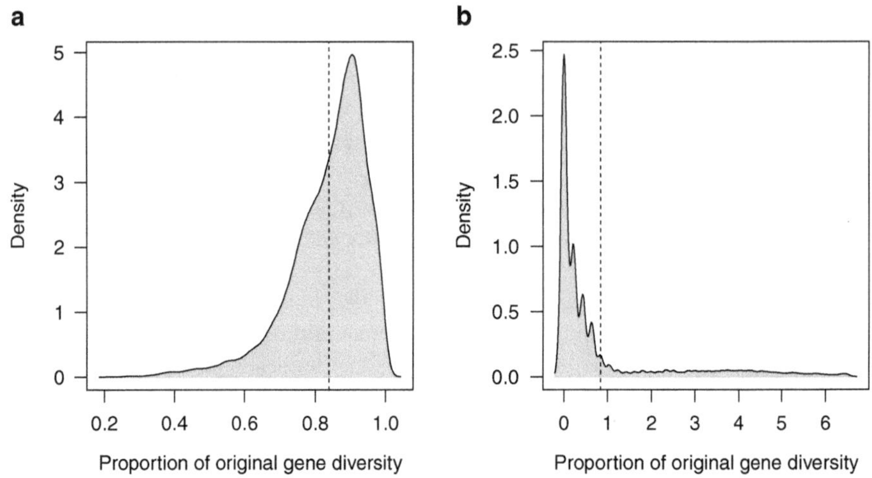

Fig. 14.2 (**a**) Density probability plot of gene diversity in the living wolverine EEP population among 10,000 simulation runs. The genome/genetic composition model involves a single–locus, two–equal–alleles model. The *dotted vertical line* is the mean value ($\bar{H}_e = 0.84$). (**b**) Density probability plot of gene diversity with single–locus, rare–allele ($p = 0.04$) model and 10,000 simulation runs. The *dotted vertical line* is the mean value ($\bar{H}_e = 0.87$)

Table 14.2 Effect of "wild" allele frequencies (p_1 to p_{16}) in a single locus model in simulations on mean and sample variances in gene diversity (H_e) in the wolverine population (31 December 2000). The values are presented as proportion of the original ("wild") genetic variation. Experiments *A* to *E* follow the ChromoFlow/GeneFlow model, experiment *F* is a gene drop model. The number of simulation runs in all experiments is 10,000

		Wild	Proportional to wild	
No.	Wild allele frequencies	H_e	H_e	σ_{H_e}
A	$p_{1-2} = 0.500$	0.500	0.84	0.034
B	$p_{1-4} = 0.250$	0.750	0.84	0.012
C	$p_{1-8} = 0.125$	0.875	0.84	0.006
D	$p_{1-16} = 0.0625$	0.9375	0.84	0.003
E	$p_1 = 0.96, p_2 = 0.04$	0.0768	0.87 (0.83)[b]	2.290 (2.13)[b]
F	$p_{1-16} = 0.0625$[a]	1.000	0.78[c]	0.000[c]

H_e gene diversity or expected heterozygosity, σ_{H_e} variance in gene diversity
[a]Unique alleles are assigned to the eight founders
[b]100,000 iterations
[c]Living wild–born individuals are not included

Figure 14.2a, b present density distributions of the proportion of original gene diversity over 10,000 runs, in the living population for models with equal allele frequencies and a rare allele ($p = 0.04$), respectively. Mean and variance for gene diversity in both simulation experiments are presented in Table 14.2 (experiments *A* and *E*).

The density plots are created in R (R Core Team 2015) and used in this section for illustration purposes only, i.e. to show the shape of the distributions of gene diversity under different genome/genetic models.

The "rare allele" simulation experiment shows a wide range in simulation values, but also a density distribution that is skewed to the left, indicating a high probability that the rare allele in this population has been lost, either due to founder effect or genetic drift (see Fig. 14.2b). The variance in this model is high, as expected from the density distribution, compared to experiment *A*, i.e. 2.290 and 0.034, respectively (see Table 14.2).

The mean value of gene diversity in the rare allele model (experiment *E*) differs from the results of experiments *A* to *D*. This indicates that 10,000 iterations is not sufficient for this particular model. A re–run with 100,000 iterations results in a more similar mean value (value between brackets in Table 14.2).

Table 14.2 shows that variances in genetic variation decline for increasing levels of original gene diversity. Experiment *D* with 16 equally distributed alleles results in the lowest variance for H_e in the living population in the group experiments *A* to *E*. This result means that a simulation with a single locus model that assumes several equally distributed alleles is adequate, even with modest numbers of iterations, to estimate the *expected* proportion of original (wild) gene diversity that has been retained.

The conclusion above also applies to the (single locus) gene drop model as implemented in GENES and PMx (Ballou et al. 2010; Lacy 1993). Experiment *F* is based on this model. An important difference with the ChromoFlow/GeneFlow model is that the founder effect is not simulated, as unique alleles are directly assigned to the founders.

GENES assesses success of the transmittance of original (wild) genetic variation into the captive population (also the default option in PMx). This means that (living) wild–born individuals are not considered as part of the living population. This explains the lower value for gene diversity in experiment *F* compared to ChromoFlow/GeneFlow experiments in which all living individuals are included.

14.5.2 *Effective Size*

Section 14.3 presented demographic models to estimate effective population size. Extreme generation overlap and non–random mating have an impact on the validity of these estimates. Gene dropping models can be used to estimate N_e from expected genetic loss between census intervals of a generation time. This allows the study of the effects of demographic trends on genetic variation.

A simulation experiment with the ChromoFlow program with a "single locus and two equal alleles" model and 1,000 simulation runs on the Nepalese red panda population was carried out. Figure 14.3 presents the expected proportion of original ("wild") gene diversity retained at census dates between 1980 and 2005.

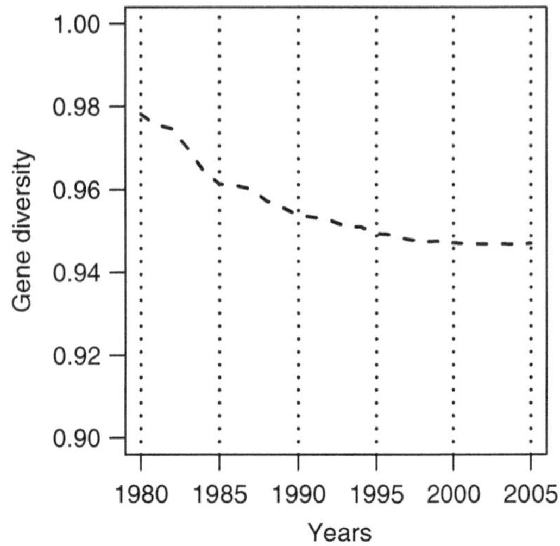

Fig. 14.3 Expected gene diversity at census dates in the Nepalese red panda population from 1980 to 2005. *Vertical bars* refer to generation intervals of 5 years. A "single locus and two equal alleles" model and 1,000 simulation runs were used

Table 14.3 Effective population size estimated from genetic loss between census intervals of 5 years in Nepalese red pandas. A "single locus and two equal allele" model and 1,000 simulation runs was used

Year	N	H_e	ΔH_e	N_e	N_e/N
1980	151	0.978	NA	29.4	0.19
1985	155	0.961	0.017	71.4	0.46
1990	173	0.954	0.007	100	0.58
1995	260	0.949	0.005	250	0.96
2000	380	0.947	0.002	∞[a]	
2005	458	0.947	0.000	NA	NA

N census size, H_e gene diversity, ΔH_e proportional loss in gene diversity, N_e effective population size, N_e/N effective size/population size ratio
[a]Effective size cannot be estimated as $\Delta H_e = 0.0$

The interval between vertical lines in this figure roughly corresponds to a generation time of 5 years in this population. The decline in gene diversity eases off over the years, with almost no loss between the 2000 and 2005 census dates. Table 14.3 presents levels of genetic variation (H_e) at these 5 year census intervals.

The effective population size N_e in this table is estimated from the "observed" proportional loss in genetic variation within a single interval ΔH_e. For example, the N_e in 1980 is estimated from the loss of 0.017 between 1980 and 1985 (see Eq. 14.1). In addition, census size (N) and N_e/N ratios are presented. Effective population size is increasing (as genetic loss per interval decreases) over the period 1980 to 2005 as a result of a growing population. The increase in N_e/N ratio indicates that a higher proportion of red pandas reproduced.

No loss in gene diversity ($\Delta H_e = 0.000$) is observed between 2000 and 2005. This means that N_e is infinitely large (Table 14.3). The zero loss could be an artefact due to sampling errors in the simulation and/or rounding errors of H_e and ΔH_e to three digits (as the number of runs is 1,000). For example, in the case that ΔH_e was 0.0005, the corresponding effective size would then be $0.5/0.0005 = 1{,}000$.

Generation overlap due to the definition of captive–born generation (see Chap. 12) can also result in minimal differences in genetic variation. This particularly applies to imports from the wild (whose genes are included in ChromoFlow estimations of census groups) and/or F1 offspring born in the period 2000–2005. These individuals will "add" genetic variation to a population which is mainly composed of older captive–born generations (see Princée 1988, 1989).

Simulation models using pedigree data enable the estimation of genetic variation (in neutral loci) that is expected to have been retained in the living population and they provide information on historic patterns of genetic loss in the population. This information can be used to evaluate which management measures need to be taken in order to maintain sufficient levels of genetic variation in the managed population.

Gene drop simulations on real–time pedigrees enable us to study the effects of complex population dynamic processes on genetic variation. In addition, chromosome–based models that involve crossing–over, such as ChromoFlow (Princée 1998), can provide a detailed insight into genetic processes.

References

Allendorf FW, Luikart G, Aitken SN (2013) Conservation and the genetics of populations, 2nd edn. Wiley–Blackwell, Chichester

Ballou JD, Foose TJ (1996) Demographic and genetic management of captive populations. In: Kleiman DG, Allen ME, Thompson KV, Lumpkin S (eds) Wild mammals in captivity: principles and techniques, 1st edn. The University of Chicago Press, Chicago, pp 263–283

Ballou JD, Lees C, Faust LJ, Long S, Lynch C, Bingaman Lackey L, Foose TJ (2010) Demographic and genetic management of captive populations. In: Kleiman DG, Thompson KV, Kirk Baer C (eds) Wild mammals in captivity: principles and techniques for zoo management, 2nd edn. The University of Chicago Press, Chicago, pp 219–252

Ballou JD, Lacy RC, Pollak JP (2011) PMx: software for demographic and genetic analysis and management of pedigreed populations (version 1.0). Technical report, Chicago Zoological Society, Brookfield

Crow JF, Kimura M (1970) An introduction to population genetics theory. Harper and Row, New York

Engen S, Lande R, Saether BE (2005) Effective size of a fluctuating age–structured population. Genetics 170(2):941–954

Felsenstein J (1971) Inbreeding and variance effective numbers in populations with overlapping generations. Genetics 681(4):581–597

Frankham R, Ballou JD, Briscoe DA (2010) Introduction to conservation genetics, 2nd edn. Cambridge University Press, Cambridge

Gilbert JP, Hammel EA (1966) Computer simulation and analysis of problems in kinship and social structure. Am Anthropol 68(1):71–93

Haldane JBS (1919) The combination of linkage values, and the calculation of distances between the loci of linked factors. J Genet 8:299 309

Kimura M, Crow JF (1963) The measurement of effective population number. Evolution 17(3):279–288

Lacy RC (1993) Genes: a computer program for the analysis of pedigrees and genetic management of populations. Software, Chicago Zoological Society, Brookfield

Lande R, Barrowclough G (1987) Effective population size and genetic variation. In: Soulé ME (ed) Viable populations for conservation. Cambridge University Press, Cambridge, pp 87–123

MacCluer JW (1967) Monte Carlo methods in human population genetics: a computer model incorporating age-specific birth and death rates. Am J Hum Genet 19(3):303–312

MacCluer JW, VandeBerg JL, Read B, Ryder OA (1986) Pedigree analysis by computer simulation. Zoo Biol 5:147–160

Princée FPG (1988) Genetic variation in the zoo population of the red panda subspecies *Ailurus fulgens fulgens*. Zoo Biol 7(3):219–231

Princée FPG (1989) Preservation of genetic variation in the red panda population. In: Glatston AR (ed) Red panda biology. SPB Academic Publishing, The Hague, pp 171–182

Princée FPG (1998) Genetic management of small animal populations in zoos and wildlife reserves: the use of genome models in the estimation of genetic variation and the effects of social structures. PhD thesis, University of Groningen, Groningen

R Core Team (2015) R: a language and environment for statistical computing. R foundation for statistical computing, Vienna, Austria. https://www.R-project.org/

Scobie P, Flesness NR (1989) Single Population Animal Record Keeping System (SPARKS), software and manual. International Species Information System, Apple Valley

Stam P (1979) Interference in genetic crossing over and chromosome mapping. Genetics 92(2):573–594

Starfield AM, Bleloch A (1991) Building models for conservation and wildlife management, 2nd edn. Burgess International Group, Edina

Wright S (1922) Coefficients of inbreeding and relationship. Am Nat 56:330–338

Wright S (1931) Evolution in Mendelian populations. Genetics 16(2):97–159

Chapter 15
Avoidance of Selection

Abstract Avoidance of selection in captive populations of wild species is discussed in the context of maintenance of original genetic variation. The complexity of identifying advantageous and disadvantageous alleles is illustrated with sickle cell haemoglobin in humans and clutch size in great tits. Consistent artificial selection for preferred wild phenotypes that may have occurred in the past is described. Unconscious selection as a result of the captive environment is also discussed. The chapter presents methods that are used in species management to minimise unconscious selection. The concept of founder representation in detecting under–represented lineages in the (living) population and identifying genetically important individuals is described. Founder representation in the European studbook population of wolverines is used as an example. It illustrates difficulties associated with over-representation in a population that was established with a small number of founders. The concept of estimating mean kinship of individuals from kinship coefficients is described. The procedure used to identify genetically important individuals is illustrated with mean kinship in male and female wolverines.

15.1 Good or Bad Genes?

Maintenance of (natural) genetic variation in captive populations of wild species is important, whether the population is part of a reintroduction programme or is to be managed long–term in captivity. These populations need to have sufficient levels of genetic variation to be able to cope with changing environments (as does any population). Environmental changes may already have occurred in the original habitat during the period in captivity. For example, the reintroduction of Przewalski's horses (*Equus przewalskii*) started in 1992, almost a century after the first founders arrived in zoos and some 30 years after the species was assumed to be extinct in the wild (van Dierendonck and Wallis de Vries 1996). Climate data for the period 1940–2006 show that the mean annual temperature in Mongolia has increased by 2.14 °C, with noticeably increased summer temperatures (Dagvadorj et al. 2009).

An underlying thought with regard to maintaining genetic variation is that original (wild) alleles in the founder population cannot simply be labelled in terms of being "advantageous" or "disadvantageous". The captive environment will

F.P.G. Princée, *Exploring Studbooks for Wildlife Management and Conservation*,
Topics in Biodiversity and Conservation 17, DOI 10.1007/978-3-319-50032-4_15

almost certainly differ from the natural environment, and, therefore, alleles that are advantageous in the wild can be disadvantageous under captive conditions, and vice versa. Moreover, it is difficult to assess which alleles in a genome (with 4,000–50,000 structural loci (Nei 1987)) are advantageous, and especially to predict their natural state at time of reintroduction. Even recessive deleterious alleles can be advantageous in the heterozygous state. A "classic" example is protection against malaria in humans who are heterozygous for the sickle haemoglobin (HbS). Although homozygous individuals for HbS will not reach adulthood, a balanced polymorphism exists in malaria infected areas (Eridani 2011).

A more important reason not to think in terms of advantage of alleles, is the fact that natural environments are not constant, but fluctuate from season to season, and from year to year. Selective forces will consequently fluctuate accordingly. Furthermore, selection acts at the level of phenotypes, which are the result of genotypes, maternal effects and environmental factors. Thus phenotype does not even refer to a single gene, but often to traits that involve tens to hundreds of genes. Methods to estimate genetic and environmental components in traits are discussed in Chap. 16.

Clutch sizes in wild bird populations have been relatively well studied and have a clear genetic component (Postma and van Noordwijk 2005; van Noordwijk et al. 1981). This trait is a good example with which to illustrate the complexity of natural selection. One would expect that selective forces would act in favour of larger clutch sizes, but long–term studies have not found such evidence (Postma and van Noordwijk 2005). For example, survival of young in large clutches of great tits (*Parus major*) is lower in "bad" years than survival of those in small clutches (Boyce and Perrins 1987). Fluctuation in environmental factors that affect food abundance is a mechanism that explains maintenance of genetic variability in clutch size in the great tit.

15.2 Selection

Captive populations are also subject to selection. Two base types of selective forces that occur in captive populations were identified by Darwin (1875):

1. Methodical artificial selection that favours specific genotypes or phenotypes (traits).
2. Unconscious selection due to selective forces of the captive environment, which includes local environmental conditions.

Methodical artificial selection is generally associated with breeding domesticated species. This type of selection favours traits that are either beneficial, e.g. milk production in dairy cattle, or perceived in terms of aesthetics, e.g. coat patterns or morphological shape in pets. Inbreeding schemes are often applied in order to increase the chance that offspring inherit the desired trait. Such breeding schemes are also applied by breeders of albino morphs of lions and tigers (AZA Welfare Committee 2011).

Artificial selection has most likely taken place in zoo populations established before population genetics was introduced in management programmes. For example, wild Sumatran tigers (*Panthera tigris sumatrae*) which were imported to The Netherlands in the 1950s were selected in favour of larger size for zoological gardens (L. de Boer, personal communication). Phenotypic selection may have taken place in a wide range of species, such as for antlers in deer or manes in African lions (*Panthera leo*). Some of these individuals may have become founders of recent zoo populations. However, this type of historical selection cannot easily be reversed.

Unconscious selection favours genotypes and phenotypes that are "adapted" to the captive environment. Since this type of selection is unnoticed it can lead to changes in allele frequencies and loss of alleles that may jeopardise success of future reintroduction programmes (Arnold 1995; Frankham 2008; Williams and Hoffman 2009).

Equalising founder representation in the population, and arranging mating combinations based on mean kinship values of individuals, are methods that are used in the management of studbook populations to minimise selection (Leus et al. 2011). These methods are discussed in the following sections.

15.3 Founder Representation

A *founder* is defined as an individual with no known genetic relationship to any other individual in the pedigree except for its descendants (Lacy 1989). In the context of captive management, founders refer to wild–born individuals and individuals with unknown parentages from other captive sources.

Equalising family sizes among founders and their descendants in subsequent generations will not only reduce random genetic drift (see Chap. 14), but will also minimise selection through equal distribution of founder genomes in the population. However, this condition is rarely met in populations of wild animals. The real distribution of founder genomes can be calculated by summation of founder representation in living individuals (Foose 1983). The *additive relationship matrix* method, as used to calculate inbreeding coefficients, is also applied to calculate founder representation in individuals (see Ballou 1983).

The European zoo population of wolverines in December 2000 consisted of 55 living individuals. The total number of founders is eight, of which four were alive in December 2000. Figure 15.1 presents the founder representation in the captive–born individuals. The horizontal line represents the equal founder representation in this population (which with eight founders is 0.125).

Wolverines did breed in zoos before an EEP was established in 1994. The starting EEP population was small and based on a few founders. This made it difficult immediately to correct effects such as unequal founder representation. For example,

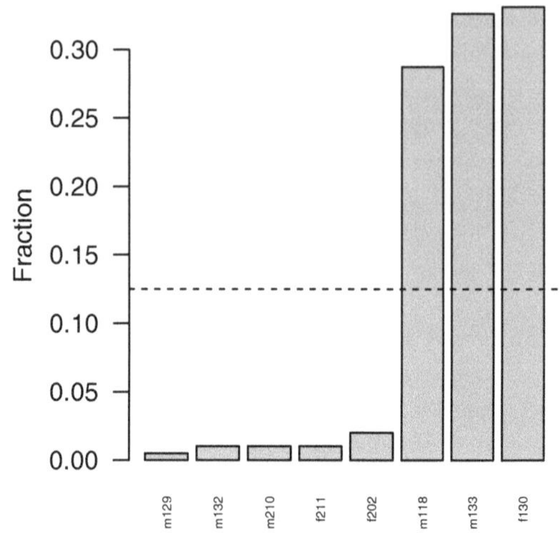

Fig. 15.1 Founder representation in the European zoo population of wolverines on 31 December 2000. Studbook IDs of individuals are on the X-axis. The *dashed horizontal line* represents equal founder representation (0.125)

three founders represent almost 95 % of the founder genomes in the year 2000 population (Fig. 15.1).

Breeding recommendations in the early years of population management would focus on equalising founder representation. This can be a complex process, depending on the previous history. For example, all living captive–born wolverine individuals in 2000 were related to the overrepresented female 130. This makes it impossible to breed any of these individuals without increasing founder representation of this founder female.

Three wild–born wolverines reproduced for the first time, increasing the genetic variation in the captive population, in 2000 (Blomqvist 2001). However, these wild–born individuals were paired with captive–born individuals. This means that their offspring represent both rare and common founder lineages. These mating combinations are not optimal from a genetic perspective, as equalising founder representation for the new founder lineage automatically increases representation of the common lineages. Combinations between the wild–born wolverines would have been preferred, but in practice this may not have been a feasible option, e.g. due to unequal sex–ratio (1 male and 2 females) or different (distant) locations.

15.4 Mean Kinship

Choosing which individuals to breed can be very complex when founder lineages are, almost inevitably, mixed over the generations. The *mean kinship* (symbol: *mk*) method was developed to select individuals that are genetically important for the population (Ballou and Lacy 1995). The concept is based on identifying those individuals that are least related to the others in the population. The genetic relatedness between two individuals i and j can be expressed as the *kinship*

Table 15.1 Mean kinship table of 10 male and female wolverines in the European zoo population (in 2000) ranked by lowest mean kinship values

Males ID	Sire	Dam	*F*	*mk*	Females ID	Sire	Dam	*F*	*mk*
210	WILD	WILD	0	0.005	220	WILD	WILD	0	0
235	202	203	0	0.1241	211	WILD	WILD	0	0.005
240	152	210	0	0.1288	202	WILD	WILD	0	0.01
157	139	118	0	0.195	185	134	132	0	0.0508
146	139	118	0	0.2025	234	211	144	0	0.1159
166	139	118	0	0.205	236	202	203	0	0.1241
226	170	166	0.1875	0.2195	130	WILD	WILD	0	0.1656
212	161	143	0.3125	0.2198	167	139	118	0	0.195
213	161	143	0.3125	0.2198	153	139	118	0	0.195

ID studbook ID, *F* inbreeding coefficient, *mk* mean kinship

coefficient (k_{ij}), which is the inbreeding coefficient of their (hypothetical) offspring. Kinship coefficients are also computed with the additive relationship matrix method (Henderson 1976; Quaas 1976).

Mean kinship (*mk*) is the mean of kinship coefficients between individual *i* and all living non–founder individuals (including itself i.e. if not a founder):

$$mk_i = \frac{\sum_{j=1}^{N} k_{ij}}{N} \tag{15.1}$$

where *N* is the number of living individuals (Ballou and Lacy 1995).

The general procedure is to rank males and females according to *mk* value (lowest first). Individuals that are unrelated to the others, such as wild–born individuals that have not reproduced, will have an $mk = 0$ and will be listed at the top.

Table 15.1 presents ranked mean kinship tables for male and female wolverines in the European zoo population in the year 2000. Wild–born female 220 had not yet reproduced in 2000 and is therefore listed at the top. Female 130, who is overrepresented, has an *mk* value of 0.1656. Offspring of the wild–born individuals 202, 210 and 211 have high *mk* values due to an overrepresented parent e.g. male 240 has a value of 0.1288.

The individuals at the top of the mean kinship list are the first to "select" for breeding. It depends on demographic management how many males and females from the list are allowed to breed. The best mating combinations are not necessarily between individuals of the same rank as they could be related e.g. siblings. Wild–born wolverine individuals can belong to the same litter (den) (Blomqvist 2012). The *MateRx* model is used by studbook keepers and population managers to create a list of recommended combinations in which inbreeding coefficients of potential offspring are taken into account (Ballou et al. 2001; Traylor Holzer 2011).

Age structure of the population is not considered in mean kinship calculations (Ballou and Lacy 1995). This means that post–reproductive individuals, who can no longer contribute genes to future generations, inflate mean kinship values of their descendants.

The solution to this problem is the *kinship value*, which is the mean kinship weighted for the reproductive values of each living individual (Ballou and Lacy 1995). Implementation of kinship values depends on the reliability of reproductive values, and thus depends on available sample sizes within a studbook (see Chaps. 7 and 9).

Mating schemes based on equalising founder representation and/or mean kinship values can (partly) prevent selection by ensuring that potential founders and descendants of underrepresented founder lineages reproduce successfully. Chapter 16 will discuss further avoidance of selection by taking quantitative traits into account.

References

Arnold SJ (1995) Monitoring quantitative genetic variation and evolution in captive populations. In: Ballou et al. (1995), pp 295–317

AZA Welfare Committee (2011) Welfare and conservation implications of intentional breeding for the expression of rare recessive alleles. Technical report, Association of Zoos and Aquariums. http://www.aza.org/assets/2332/aza_white_paper_inbreeding_for_rare_alleles-_18_jan_2012.pdf

Ballou JD (1983) Calculating inbreeding coefficients from pedigrees. In: Schonewald-Cox et al. (1983), pp 509–520

Ballou JD, Lacy RC (1995) Identifying genetically important individuals for management of genetic variation in pedigreed populations. In: Ballou et al. (1995), pp 76–111

Ballou JD, Gilpin M, Foose TJ (eds) (1995) Population management for survival and recovery: analytical methods and strategies in small population conservation. Methods and cases in conservation science series. Columbia University Press, New York

Ballou J, Earnhardt JM, Thompson SD (2001) MateRx: genetic management software, 1st edn. National Zoological Park, Washington DC/Lincoln Park Zoo, Chicago

Blomqvist L (2001) Management of captive wolverines *Gulo g. gulo* in Europe: studbook and guidelines to husbandry, vol 2. Nordic Ark, Hunnebostrand

Blomqvist L (2012) Husbandry manual for captive wolverine, *Gulo gulo*, 2012. Nordense Ark Foundation, Hunnebostrand

Boyce MS, Perrins CM (1987) Optimizing great tit clutch size in a fluctuating environment. Ecology 68(1):142–153

Dagvadorj D, Natsagdorj L, Dorjpurev J, Namkhainyam B (2009) Mongolia: assesment report on climate change 2009. Technical report, Ministry of Environment, Nature and Tourism, Ulaanbaatar

Darwin C (1875) The variation of animals and plants under domestication, vol II, 2nd edn. John Murray, London

Eridani S (2011) Sickle cell protection from malaria: a review. Hematol Rev 3(3):72–77

Foose TJ (1983) The relevance of captive populations to the conservation of biotic diversity. In: Schonewald-Cox et al. (1983), pp 374–413

Frankham R (2008) Genetic adaptation to captivity in species conservation programs. Mol Ecol 17(1):325–333

Henderson CR (1976) A simple method for computing the inverse of a numerator relationship matrix used in prediction of breeding values. Biometrics 32:69–83

Lacy RC (1989) Analysis of founder representation in pedigrees: founder equivalents and founder genome equivalents. Zoo Biol 8:111–123

Leus K, Traylor–Holzer K, Lacy RC (2011) Genetic and demographic population management in zoos and aquariums: recent developments, future challenges and opportunties for scientific research. Int Zoo Yearb 45:213–225

Nei M (1987) Molecular evolutionary genetics. Columbia University Press, New York

Postma E, van Noordwijk AJ (2005) Genetic variation for clutch size in natural populations of birds from a reaction norm perspective. Ecology 86(6):2344–2357

Quaas RL (1976) Computing the diagonal elements and inverse of a large numerator relationship matrix. Biometrics 32:949–953

Schonewald-Cox CM, Chambers SM, MacBryde B, Thomas L (eds) (1983) Genetics and conservation: a reference for managing wild animal and plant populations. Biological conservation, Benjamin/Cummings, Menlo Park

Traylor–Holzer K (ed) (2011) PMx users manual, Version 1.0. IUCN SSC Conservation Breeding Specialist Group, Apple Valley

van Dierendonck MC, Wallis de Vries MF (1996) Ungulate reintroductions: experiences with the Takhi or Przewalski horse (*Equus ferus przewalskii*) in Mongolia. Conserv Biol 10(3):728–740

van Noordwijk AJ, van Balen JH, Scharloo W (1981) Genetic and environmental variation in clutch size of the great tit (*Parus major*). Neth J Zool 31(2):342–372

Williams SE, Hoffman EA (2009) Minimizing genetic adaptation in captive breeding programs: a review. Biol Conserv 142(11):2388–2400

Chapter 16
Quantitative Genetics

Abstract Monitoring phenotypic variation in natural history traits that can be directly obtained from studbook data is explored in this chapter. Phenotypic variances in lifespan, age at first breeding, inter–birth interval, litter size, breeding season and fitness, as observed in the captive populations of Chinese and Nepalese red pandas, are explored. The components of phenotypic variance and concepts of heritability and repeatability are described. Repeatability in different traits of red crowned cranes, red pandas and snow leopards is presented and discussed. Mid–parent and single parent regressions to estimate heritability (h^2) are described and illustrated with parturition date in red panda subspecies. Methods to handle unequal family sizes in regression and to adjust for assortative matings are explained and illustrated with the trait "fitness" in red pandas. Assumptions in linear regression regarding independent data and normal distribution of trait data are discussed, as are effects of outliers on results. The Residual or Restricted Maximum Likelihood (REML) and Markov chain Monte Carlo (MCMC) implementations of the "animal model" to estimate heritability are briefly described. Parturition date in Nepalese red pandas is used to demonstrate these methods. The last section explores the use of estimated breeding values (EBVs) in monitoring phenotypic variation. Litter size in maternal generation groups of African wild dogs is used as an example.

16.1 Phenotypic Traits

Genetic management of captive populations of wild species is predominantly based on the *neutral–mutation–random–drift theory of molecular evolution* (see Crow and Kimura 1970; Kimura 1968). Mutation does *not* result in allelic variants that are more beneficial than others, according to the neutral theory. Fixation of alleles (loss of genetic variation) is entirely due to random genetic drift in an idealised population.

Minimising genetic drift by increasing effective population size will, in effect, primarily maintain genetic variation in (selectively) neutral genes. However, this does not mean that genetic variation at loci involved in selection directed to adaptation to captivity is lost. Identification of genetically important individuals

F.P.G. Princée, *Exploring Studbooks for Wildlife Management and Conservation*, Topics in Biodiversity and Conservation 17, DOI 10.1007/978-3-319-50032-4_16

based on mean kinship values minimises (unconscious) selection between families (see Chap. 15). However, (unconscious) selection among siblings within families remains (e.g. Frankham 2008).

Selection acts at phenotypes of traits, which are the combined product of genotypes on single or often multiple genes (*polygenic*), of maternal effects, interaction between alleles and genes, and environmental factors. The genes involved in these traits are unknown, more often than not. This complexity makes it almost impossible to assess genetic variation at the level of individual loci that are involved in a specific trait. In other words, the genetic composition of a trait is often a "black box".

The field of *quantitative genetics* describes and unravels the "black box" of traits. It is likely that animal and plant breeders, in the era before Darwin and Mendel, used concepts of quantitative genetics; for example, by selecting specific bulls and/or cows to improve meat and milk production in herds. However, a more fundamental approach, describing responses to natural selection, refers to the work of R.A. Fisher and S. Wright in the early 1900s (Fisher 1918; Wright 1921b). Understanding the process of selection on traits is essential in order to minimise selection in captive populations of endangered species.

Although the importance of quantitative genetics for (genetic) management of zoo populations is recognised (e.g. Arnold 1995; Frankham 1999; Lande 1995; McDougall et al. 2006), this field of genetics is, so far, only sporadically used in species management programmes. Examples include the trait juvenile survival in Cuvier's gazelle (*Gazella cuvieri*) (Ibáñez et al. 2014) and the traits courtship display, ejaculate size, clutch size and body mass per year in Houbara bustard (*Chlamydotis undulata*) (Chargé et al. 2014). However, the potential for research in quantitative genetics on captive populations is realised (Pelletier et al. 2009). For example, the ISIS database was used to study heritability of longevity in 14 mammal and 8 bird species (Ricklefs and Cadena 2008).

Data on morphological and physiological traits are not registered as standard in studbooks. However, various life history traits can readily be studied without the need for additional data collection. Chapter 3 showed how basic studbook data, i.e. sex, parentages and dates of birth and death, are used to analyse life history traits. The sample variance of the mean as estimated in these traits is the *phenotypic variance* (symbol: V_P).

Lifespan, age at first breeding, litter size, birth season and inter–birth interval are used as examples of traits in this chapter. Additionally the trait "fitness", measured as the total number of offspring produced during an individual's life–time, is included as an example (Kruuk et al. 2000).

This chapter provides an introduction to the role quantitative genetics can play in (genetic) management of studbook populations. The red panda studbook is used as the main example to illustrate methods to estimate heritability.

16.2 Phenotypic Variance

Phenotypic variances (V_P) of life history traits in studbook populations are, as mentioned in the previous section, sample variances of the mean trait values. For example, the phenotypic variance of the trait "litter size" is the variance of mean litter size, or the phenotypic variance of "reproductive lifespan" is the variance of mean age at reproduction.

Interpretation of sample variances, for example whether small or large, is not that straightforward, especially when comparing populations or different traits (measures of unit). Coefficients of variation, which are named *phenotypic coefficients of variation* (symbol: CV_P), are unit–free and dimension–free, and thus provide a measure of relative variability (Houle 1992).

Table 16.1 presents V_P and CV_P values (in %) for life history traits in Nepalese (*fulgens*) and Chinese (*styani*) red panda subspecies. This table illustrates the differences in magnitude of phenotypic variances between traits. For example, variances of litter size and female age at first breeding in Nepalese red pandas are 0.3983 and 379,237, respectively; yet, the CV_P values of both traits are rather

Table 16.1 Phenotypic variance of life history traits in the global studbook population of red panda subspecies

Taxon	Sex	Trait	Unit	N_{obs}	Mean	V_P	CV_P
Fulgens	Male	Lifespan	Day	658	1,791	4,085,186	112.9
Fulgens	Female	Lifespan	Day	683	2,039	3,952,307	97.5
Styani	Male	Lifespan	Day	224	2,513	6,076,186	98.2
Styani	Female	Lifespan	Day	210	2,424	6,195,756	102.8
Fulgens	Male	Age at first breeding	Day	304	1,473	699,948	56.9
Fulgens	Female	Age at first breeding	Day	349	1,275	379,237	48.7
Styani	Male	Age at first breeding	Day	101	1,709	841,626	53.8
Styani	Female	Age at first breeding	Day	117	1,281	269,533	40.6
Fulgens	N/A	Inter–birth interval	Day	238	459.5	56,642	51.8
Styani	N/A	Inter–birth interval	Day	102	508.1	86,997	58.1
Fulgens[a]	N/A	Parturition date	Day number	1026	177	197.6	7.9
Styani	N/A	Parturition date	Day number	468	180.4	235.9	8.5
Fulgens	N/A	Litter size	Offspring	1228	1.714	0.3983	36.8
Styani	N/A	Litter size	Offspring	276	1.686	0.4381	39.3
Fulgens	Male	Fitness	Offspring	685	2.261	18.56	192.4
Fulgens	Female	Fitness	Offspring	683	2.261	15.35	173.4
Styani	Male	Fitness	Offspring	224	1.451	11.56	234.6
Styani	Female	Fitness	Offspring	210	1.643	11.66	208.1

N_{obs} number of observations, V_P phenotypic variance, CV_P phenotypic coefficient of variation (in %)

[a]Northern hemisphere locations in Europe and North America

similar. Furthermore, these data illustrate the differences in phenotypic variances between sexes and populations (in this case subspecies).

Phenotypic variances of parturition, expressed as "day of the year", in both Chinese and Nepalese red pandas at northern hemisphere zoo locations, are relatively low compared to life-history traits i.e. CV_Ps are around 8 %.

Variances of lifespan and fitness are high in both subspecies and sexes. For example, CV_P values of fitness of over 200 % (which means that the standard deviation is more than twice the mean value) can be observed for Chinese red pandas (see Table 16.1).

16.3 Variance Components

The phenotypic variance (V_P) can be partitioned in variance components (Falconer 1960). Since we are interested in the genetic contribution to phenotypic variance, the basic partition is in the *genetic variance* (symbol: V_G) and the *environmental variance* (symbol: V_E) (Falconer 1960):

$$V_P = V_G + V_E \tag{16.1}$$

Genetic variance (V_G) reflects genetic differences between individuals which are the end result of multiple loci, including interactions between loci. This *variance* should not be confused with measures of genetic *variation* such as gene diversity. Genetic variance is difficult to estimate from observations, and is partitioned into *additive genetic variance* (symbol: V_A); *dominance variance* (symbol: V_D), which results from interactions between alleles; and *epistatic variance* (symbol: V_I), which results from interactions between loci (Falconer 1960):

$$V_G = V_A + V_D + V_I \tag{16.2}$$

Both dominance (V_D) and epistatic (V_I) variances are difficult to estimate without specific breeding experiments. Therefore animal breeders and ecologists focus on estimating the additive genetic variance (V_A) (Wilson et al. 2010).

When repeated measurements of the same individual are available, e.g. litter sizes and inter–birth intervals, the environmental variation can be partitioned into the *general or permanent environmental variance* (symbol: V_{Eg}) and the *specialised environmental variance* (symbol: V_{Es}) which is localised (Falconer 1960):

$$V_E = V_{Eg} + V_{Es} \tag{16.3}$$

The permanent environmental variance (V_{Eg}) is the *between–individual variance* component that is permanent across repeated measures. The specialised environmental variance (V_{Es}) is the *within–individual variance* due to temporary or localised circumstances (Falconer 1960).

Several components that make up phenotypic variance (V_P) have been introduced in this section. Equation16.4 presents the main components:

$$V_P = V_A + V_D + V_I + V_{Eg} + V_{Es} \tag{16.4}$$

The environmental variance can be further partitioned into random factors that have been identified e.g. latitude or year of birth. These phenotypic variance models are discussed in Sect. 16.9.

16.4 Heritability

The genetic contribution to phenotypic variance can be expressed as a proportion named *broad–sense heritability* (symbol: H^2). This proportion is estimated as:

$$H^2 = \frac{V_G}{V_P} \tag{16.5}$$

where V_G is the genetic variance and V_P the (total) phenotypic variance.

Broad–sense heritability (H^2) is not of great practical use as V_G cannot be easily estimated. Instead a measure of resemblance between related individuals, e.g. parent and offspring, named *narrow–sense heritability* (symbol: h^2) is more commonly estimated:

$$h^2 = \frac{V_A}{V_P} \tag{16.6}$$

where V_A is the additional genetic variance and V_P the (total) phenotypic variance. The term heritability generally refers to narrow–sense heritability (h^2), and this is how it will be used here.

16.5 Repeatability

The upper limit of broad–sense heritability (H^2) can be obtained by estimating *repeatability* (symbol: r_P) between repeated measures of the same individual (Falconer 1960):

$$r_P = \frac{V_G + V_{Eg}}{V_P} \tag{16.7}$$

where V_G is the genetic variance, V_{Eg} is the permanent environmental variance (see 16.3) and V_P is the phenotypic variance.

Table 16.2 Repeatability of life history traits in females, in the International studbook populations of red pandas and snow leopards, and the European population of red cranes

Taxon	Trait	N	k	r_P	95 % CI
A. f. fulgens	Inter–birth interval	263	2.8	0.20	[0.12, 0.28]
A. f. styani	Inter–birth interval	102	2.6	0.34	[0.21, 0.48]
U. uncia	Inter–birth interval	124	1.7	0.29	[0.08, 0.47]
G. japonensis	Inter–clutch interval	55	4.7	0.30	[0.18, 0.44]
A. f. fulgens[a]	Parturition date	302	3.4	0.35	[0.29, 0.42]
A. f. styani	Parturition date	146	3.2	0.53	[0.44, 0.62]
U. uncia[a]	Parturition date	266	2.5	0.00	[−0.09,0.10]
G. japonensis	Hatch date	72	4.5	0.04	[−0.03,0.15]
A. f. fulgens	Litter size	370	3.3	0.17	[0.11, 0.24]
A. f. styani	Litter size	146	3.2	0.13	[0.03, 0.23]
U. uncia	Litter size	272	2.5	0.06	[−0.03,0.15]
G. japonensis	Clutch size	147	4.5	0.06	[0.00, 0.14]

N number of individuals, k mean observations per individual, r_P repeatability, *CI* confidence interval

[a]North America and Europe

Repeatability is the proportion of total variance in multiple measurements of a trait, e.g. litter size, that is due to differences between individuals (Dohm 2002). The range of r_P is 0–1. Repeatability (r_P) is the intra–class correlation coefficient that is estimated from between–individual and within–individual variances in one–way ANOVAs (Lessells and Boag 1987; Sokal and Rohlf 2012).

ANOVAs can be carried out separately on each sex. This method requires minimal pedigree data when the trait is related to reproduction (as in the following examples), e.g. dates of birth of offspring (clutches/litters) are required when analysing seasonality and inter–birth interval.

Table 16.2 presents repeatability values for inter–birth interval, day of birth/hatching and litter size in red pandas, snow leopards and red crowned cranes. The analyses were restricted to observations in females. The R package *ICC* (Wolak et al. 2012) was used to estimate r_P and 95 % confidence intervals.

Values for repeatabilities in Table 16.2 range from 0.00 to 0.53, see parturition date for snow leopards and Chinese red pandas. The lower confidence limit for snow leopards is negative, which can occur in ANOVA–based repeatabilities. Negative values are considered as zero (Nakagawa and Schielzeth 2010). An r_P of zero, e.g. parturition date in snow leopards, means that the variance between individuals does not contribute to the total phenotypic variance (Eq. 16.7). A value of $r_P = 0.53$ means that the variance among individuals contributes 53 % to the total phenotypic variance.

Repeatability is a rough upper limit of broad–sense (and therefore narrow–sense) heritability, and it can be an underestimate, e.g. in the case of interaction between genotype and environment or maternal/paternal effects (Dohm 2002).

The "standard" ANOVA, as used in this section, is based on a linear model in which it is assumed that the (phenotypic) data follow a normal (Gaussian) distribution. This assumption does not hold for many traits, such as those with discrete data e.g. litter size or day of birth (see Nakagawa and Schielzeth 2010). Furthermore, pseudoreplication such as litter, maternal and paternal effects, are not considered in the basic linear model.

ANOVAs that are based on generalised linear mixed models (*GLMM*) can be used to estimate repeatability for non–Gaussian traits and/or traits that are subject to pseudoreplication (Nakagawa and Schielzeth 2010). The R package *rptR* provides different methods, including GLMM, for estimating repeatability (Schielzeth and Nakagawa 2013). These methods will be discussed, in conjunction with heritability, in a later section.

16.6 Parent–Offspring Regression

Heritability can be estimated from the regression of trait values of offspring on parents. It is the regression coefficient (b) when mid–parent values (i.e. the mean of values for sire and dam) are used, or $2 \times b$ when regression is based on single parents e.g. mother–daughter (Falconer 1960). The data on offspring are grouped per family/single parent.

The use of mid–parent values is not strictly valid when variances between the sexes differ. Single parent regressions, i.e. mother–daughter and father–son, are preferred (Falconer 1960). This section presents both mid–parent and single parent regressions to illustrate potential differences.

16.6.1 Example: Parturition Date

The relatively high values for repeatability (r_p) of parturition date in both red panda subspecies, compared to other traits/species (Table 16.2), make this trait/species a good candidate to explore methods to assess heritability.

Table 16.3 presents the results of mid–parent and single parent regression for parturition date in both red panda subspecies. The regression coefficient (b) with standard error (se), results of the t-test on the regression coefficient (degrees of freedom (df), probability (p), and heritability (h^2) are presented in this table. The adjusted heritability (h^2_{adj}) will be discussed in Sect. 16.6.4.

The regression coefficients (b) in parent–offspring and father–son regression in the Nepalese red pandas (*fulgens*) do not differ from zero (null hypothesis), i.e. $p > 0.05$ (Table 16.3). The "zero" heritability is interpreted that the additive genetic variance (V_A) does not contribute to the phenotypic variance in parturition date (see Eq. 16.6).

Table 16.3 Heritability of parturition date in the International studbook population of red panda subspecies based on linear regression with grouped offspring data and weighted for family size. Fulgens: North America and Europe

Taxon	Regression model	b	se	df	p	h^2	h^2_{adj} [a]
Fulgens	Parent–offspring	0.10	0.06	206	0.073	N/A	
Styani	Parent–offspring	0.42	0.07	98	< 0.001	0.42	
Fulgens	Mother–daughter	0.18	0.07	155	0.011	0.36	0.06
Styani	Mother–daughter	0.44	0.09	62	< 0.001	0.88	0.11
Fulgens	Father–son	−0.07	0.08	143	0.37	N/A	
Styani	Father–son	0.32	0.12	65	0.0091	0.62	0.08

b regression coefficient, *se* standard error, *df* degrees of freedom, *p* probability of *t* test statistic, h^2 heritability, h^2_{adj} heritability adjusted for phenotypic correlation
[a]Phenotypic correlations in *A. f. fulgens* and *A. f. styani* are 0.83 and 0.88, respectively

However the null hypothesis that the regression coefficient does not differ from zero, is rejected in mother–daughter regression in the *fulgens* subspecies ($b = 0.18$, $p = 0.011$). Heritability (h^2) is 0.36, which is not much different from the estimated repeatability (r_p) for this trait/subpecies (see Table 16.2). This means that 36 % of the observed phenotypic variance in this trait is caused by additive genetic variance (see Eq. 16.6).

Linear regression models indicate that heritability for parturition date in the *styani* subspecies is high, i.e. between 0.42 and 0.88 (Table 16.3). This means that up to 88 % of the phenotypic variance is attributed to additive genetic variance, according to the mother–daughter linear regression model.

Standard errors and critical values of the Student's *t* distribution can be used to estimate confidence limits to regression coefficients, and thus heritability. For example, 95 % confidence limits for mother–daughter regression in *styani*, with $t_{0.05[62]} \approx 2.0$ and $se = 0.09$, are 0.44 ± 0.18. This means that confidence limits for heritability, which is $2b$ in single parent regression, are 0.52 and 1.0 (the upper boundary for h^2).

Visualising data by plotting phenotypes of offspring (either grouped or not) against parents is recommended, as it can be helpful in evaluating regression results. Figure 16.1a presents results of a weighted mid–parent–offspring regression on parturition date (in days) in the International studbook population of Chinese red pandas. Mean values of offspring are regressed against mid–parent values. The visible trend is supported with a statistically significant ($p < 0.001$) regression coefficient, and heritability (h^2) of 0.42 (Table 16.3).

Interpretation of heritability is subject to several misconceptions (Postma 2014; Visscher et al. 2008). An important misconception is to interpret h^2 as the proportion of a trait that is determined genetically. That is incorrect. A value of $h^2 > 0$ means that the trait has a genetic component, but does not provide information about the number of genes involved or which proportion of the trait is genetically determined. A heritability of zero ($h^2 = 0$) does *not* mean that a trait has no genetic component. It only indicates that the phenotypic variance is not caused by genetic variance.

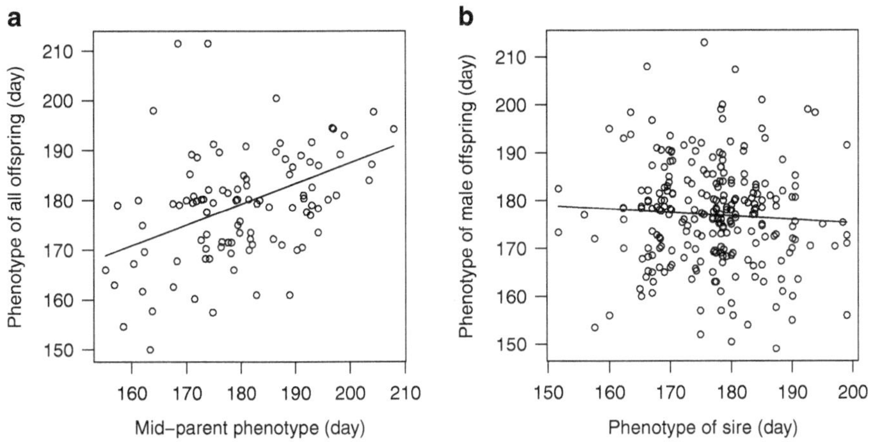

Fig. 16.1 (**a**) Weighted mid–parent–offspring regression of the trait "parturition date" in the International studbook population of Chinese red pandas. Offspring are grouped. (**b**) Father–son regression of the trait "parturition date" in the Nepalese red pandas. Offspring are not grouped

Figure 16.1b shows phenotypic data of parturition date in Nepalese red pandas between (ungrouped) father–son pairs. The distribution pattern of the data does not provide a strong feel for a linear relation, which is supported by linear regression, i.e. $p = 0.37$ (see Table 16.3).

Heritability of parturition date in the Chinese red panda is higher in mother–daughter regression analyses than in father–son and mid–parent regression. A preliminary conclusion is that variance in parturition date has a genetic component. However, the results require more detailed interpretation e.g. to explain the difference between single–parent and mid–parent regression (see Sect. 16.6.4).

16.6.2 Example: Fitness

Fitness is not only an important life history trait, but it shows high levels of phenotypic coefficients of variance (CV_P) compared to other traits in red pandas as well (see Table 16.1). This makes it an interesting trait for studying heritability. Table16.4 presents results of mid–parent and single parent regressions in Nepalese red pandas. The adjusted heritability (symbol: h^2_{adj}) in this table is explained in Sect. 16.6.4.

Regression analyses involving female offspring, i.e. parent–daughter, mother–daughter and father–daughter, did not result in statistically significant heritabilities ($p > 0.05$; Table 16.4). Fitness in male offspring, however, more resembles fitness in the mother than in the father, i.e. h^2 is 0.30 and 0.14, respectively.

Table 16.4 Heritability of fitness in the International studbook population of Nepalese red pandas based on linear regression with grouped offspring data and weighted for family size

Regression model	b	se	df	p	h^2	h^2_{adj} [a]
Parent–offspring	0.09	0.03	263	0.001	0.09	
Parent–daughter	0.03	0.04	187	0.873	N/A	
Parent–son	0.13	0.04	200	0.002	0.26	
Mother–daughter	0.01	0.04	220	0.871	N/A	
Mother–son	0.15	0.04	228	< 0.001	0.30	0.11
Father–son	0.07	0.03	223	0.026	0.14	0.05
Father–daughter	0.06	0.03	210	0.072	N/A	

b regression coefficient, *se* standard error, *df* degrees of freedom, *p* probability of *t* test statistic, h^2 heritability, h^2_{adj} heritability adjusted for phenotypic correlation
[a]Phenotypic correlation between parents is 0.62

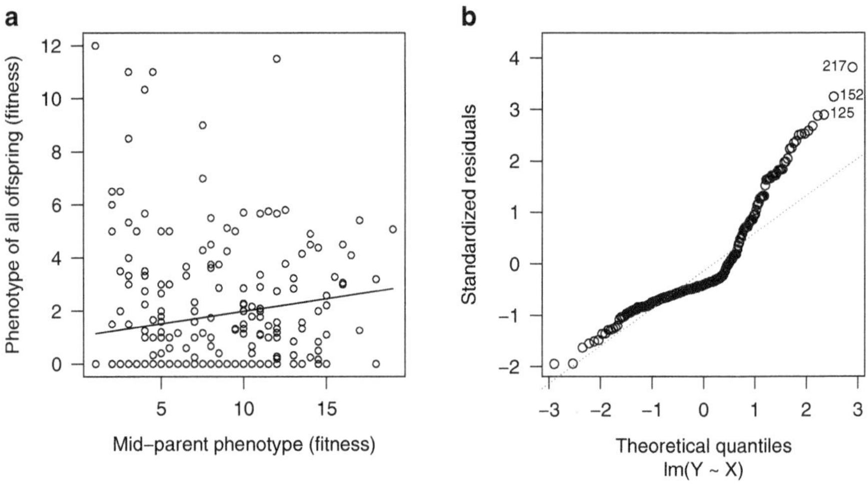

Fig. 16.2 (**a**) Weighted mid–parent–offspring regression of the trait "fitness" in the International studbook population of Nepalese red pandas. Offspring are grouped. (**b**) Theoretical and actual standardised residuals

Figure 16.2a presents results of mid–parent–offspring regression on fitness in Nepalese red pandas. The regression line reflects the low (but significant) heritability ($h^2 = b = 0.09$, $p = 0.001$) for this trait.

Interpretation of the differences in heritability of fitness between females and males is complex as it encompasses an entire individual's lifespan. Again, a non–significant regression coefficient (and heritability) does *not* mean that a trait has no genetic component. Non–genetic factors seem to determine most of the phenotypic variance in fitness of females. Fitness in these analyses refers to total number of progeny produced during an individual's lifespan. One could imagine that non–genetic factors related to "giving birth" can affect survival and further reproduction in females.

16.6.3 Weighted Regression

Parents with a single offspring provide less information on (cor)relations in phenotypes between parents and offspring than parents with large numbers of progeny. This section presents results of parent–offspring regression that is weighted against family size.

Although this is not the best weighting method, it has the advantage that a basic linear model can be applied. The preferred weighting method is based on the regression coefficient itself (Kempthorne and Tandon 1953; Reeve 1955), and requires an iterative re–weighting procedure. When using weighted family size it is recommended to at least compare results with unweighted regression, as the "true" value will be somewhere in between (Arnold 1994). For example, heritability of parturition date in Chinese red pandas is 0.38 and 0.42 for unweighted and weighted mid–parent offspring regression, respectively. This means that between 38 % and 42 % of the observed variance in parturition date is caused by additive genetic variance.

16.6.4 Assortative Mating and Artefacts

Assortative matings, i.e. matings between partners with similar phenotypes that occur more frequently than expected from random mating, increase the regression coefficient (b) in single–parent–offspring regression with a factor $(1 + r_{pp})$ (Wright 1921a). The parameter r_{pp} is the *phenotypic correlation between parents (mates)*. This suggests that the heritability estimated from this regression model as $h^2 = 2b$ is an overestimate (i.e. when the correlation is statistically significant). Artificial selection by pairing animals with the same phenotype has the same effect on heritability.

The phenotypic correlation is estimated as the *Pearson product–moment correlation* between phenotypes of the parents. The adjusted heritability h^2_{adj} is then estimated as (Weis 2005):

$$h^2_{adj} = 2b(1 - r_{pp}) \tag{16.8}$$

Table 16.3 includes adjusted values for heritability for single–parent–offspring regression of the trait "parturition date". The phenotypic correlation between parents in the Chinese and Nepalese red panda populations are 0.88 and 0.83, respectively (both correlations are statistically significant, i.e. $p = 0$). These high correlations reduce the heritability considerably, e.g. from 0.88 to 0.11 in mother–daughter regression in the Chinese red panda (Table 16.3).

It is not surprising that breeding partners share the same parturition date more often than would be expected under random mating, especially in the context of species management programmes. First, red pandas in zoos are managed according to mean kinship values and avoidance of inbreeding. This means that the high phenotypic correlation can not be explained by assortative mating or artificial selection. Some 80 % of established breeding pairs in red pandas do not change year on year, and are in effect monogamous. Partners in these pairs share the same litters and, consequently, have the same phenotypic value (mean parturition date). The fact that the phenotypic correlation for this specific trait is a kind of artefact, does not change the implication that heritability needs to be adjusted.

Second, the phenotypic trait "fitness" also embeds an artefact as partners in monogamous breeding pairs share the same number of offspring produced during their life–time. This explains the relatively high phenotypic correlation observed in the trait fitness in Nepalese red pandas ($r_{pp} = 0.62, p = 0$). This means that heritability of fitness (h^2) in single–parent regressions are overestimates and need to be adjusted accordingly (Eq. 16.8). These adjustments result in reducing heritabilities from 0.30 to 0.11 and 0.14 to 0.05 in mother–son and father–son regressions, respectively (see Table 16.4).

Section 16.9 will present the *animal model* in which the results are not biased by effects of assortative mating or selection (Kruuk 2004).

16.7 Assumptions of Linear Regression

The previous section presented heritability values for the traits parturition date and fitness, as estimated from standard linear parent–offspring regression. The underlying assumptions of linear regression (see Sokal and Rohlf 2012) were "conveniently" ignored. Violation of these assumptions, however, can lead to incorrect results. Statistical software, e.g. the R package *car* (Fox and Weisberg 2011), provides tools to test whether parent–offspring regression meets these assumptions. Here we focus on two important assumptions concerning the dependent variable "phenotype of offspring": that the data are (1) independent and (2) normally distributed for any given (mid–)parent phenotype.

16.7.1 Independent Data

The first assumption is related to pseudoreplication (Hurlbert 1984). The *maternal effect*, which refers to the phenotype of the mother affecting the phenotype of her offspring, and the *brood or litter effect*, a combination of maternal and litter–specific environmental effects, are examples (Kruuk and Hadfield 2007; McAdam et al. 2014). These effects are avoided in parent–offspring regression by using the mean trait values of offspring (Åkesson et al. 2008). Linear mixed regression (*LMM*) and generalised estimating equations (*GEE*) can be used to model maternal and litter effects as a random effect (see also Chaps. 10 and 13).

16.7.2 Normal Distribution

The second assumption refers to the distribution of errors or residuals, i.e. the differences between original and fitted values, in linear regression. Figure 16.2a shows weighted mid–parent–offspring regression of the trait fitness in Nepalese red pandas. Although the regression coefficient is significant ($b = 0.09$, $p = 0.001$), the phenotypic data of offspring per mid–parent value are "widely" spread. Figure 16.2b presents a Q–Q plot between the theoretical (normal distribution) and actual (standarised) residuals for this regression. The points would lie on the dashed line when the actual residuals are distributed normally. The right half of this plot clearly shows that fitness errors are not normally distributed. In addition (e.g. when in doubt), the *Shapiro–Wilk normality test* could be applied to the residual data (Sokal and Rohlf 2012). The default implementation in R (R Core Team 2015) of this test was used, and confirms the conclusion from the Q–Q plot, that the residuals are not normally distributed ($W = 0.76$, $p < 0.001$).

Results of parent–offspring regressions based on the (standard) linear model are not valid when the residuals are not normally distributed, as in the case of the trait "fitness". Potential solutions to the problem of non–normality are to transform phenotypic data, for example to logarithmic or square root, to approximate a normal distribution (of the residuals) in standard linear regression (Sokal and Rohlf 2012). Alternatively one can apply a generalised linear model (*GLM*) on data that (are assumed to) follow other known distributions e.g. binomial or Poisson. Generalised linear mixed models (*GLMM*) support random effects and thus can combine non–normality and pseudoreplication. These more advanced regression models are discussed in the following section.

Transformation may be necessary when the data do not fit one of the (known) distributions. Trying different transformations until the most "significant" regression coefficient is found, is not a sound method. Instead, select a transformation that is known to be appropriate for the data, in advance (McComb et al. 2010). Alternatively, the *Box–Cox power transformation* (Box and Cox 1964) can be applied to select an appropriate transformation (see also Sokal and Rohlf 2012).

Non–normal distribution of phenotypic data can affect results of linear regression, potentially leading to wrongly accepting or rejecting the null hypothesis that the regression coefficient (thus heritability) is zero. Transformation of data or using generalised regression models need to be applied when non–normality is expected.

16.7.3 Outliers

Linear regression is sensitive to outliers. Outlying observations can have a large effect on the regression coefficient and inflate its variance (Vittinghoff et al. 2005). For example, the top–left data point in Fig. 16.2a refers to a mid–parent fitness of 0 and a mean fitness of their offspring of 12. The Q–Q plot as implemented in

R (R Core Team 2015) marks the three most extreme residuals with labels that refer to the original data. Figure 16.2b presents three outliers in the fitness data.

Outliers that affect the regression coefficient are considered *influential*. It is good practice to check whether these data points are errors of measurement. Removal of erroneous data would be justified (see e.g. Keller et al. 2001). Whenever extreme data show to be valid observations, one needs to test whether they should be considered as influential outliers (see Crawley 2007; Fox and Weisberg 2011).

16.8 Estimated Breeding Values

Livestock and plant breeders ideally select individuals for breeding on the basis of expected phenotypes that are above the population average for a given trait in future offspring. The measure *Estimated Breeding Value (EBV)* of an individual is used in this selection process. Although selection is to be avoided in captive populations of wild species, the underlying method(s) to estimate breeding values are of interest. This is particularly true of methods that have been developed to estimate heritability (h^2) in livestock breeding.

This section briefly introduces the concept of estimated breeding values, while related methods to estimate heritability will be discussed in the next section. Section 16.10 returns to estimated breeding values as a tool to monitor phenotypic variation and to detect unconscious selection.

The method used to estimate breeding values depends on the available information on phenotypes. The basic method is based on a single record per individual:

$$EBV_i = h^2(Ph_i - \bar{Ph}) \tag{16.9}$$

where h^2 is the heritability; Ph_i is the individuals's phenotypic value; and $\bar{Ph}$ is the mean phenotypic value in the population (based on Mrode 2000).

The predictive value of EBV_i depends on the available information on an individual's phenotype *and* the accuracy of the estimated h^2 for a trait. Accuracy increases when the phenotype is based on the average of repeated records. This increase slows down after 4–5 records (Mrode 2000).

In the case that a trait can only be measured in one of the sexes, breeding values of the opposite sex are based on the mean phenotypes of progeny. A classical livestock example is selection of a bull based on the milk production in his daughters. But what if the bull has not yet reproduced? It would seem impossible to estimate its breeding value. This problem of missing phenotypic data applies to any trait. In such cases the breeding value is estimated as the mean value of both parents (the bull's parents in this case) (Mrode 2000).

16.9 Animal Model

The accuracy of breeding values *and* heritability can be improved by using all available information in a pedigree. C.R. Henderson developed a linear mixed model using *Best Linear Unbiased Prediction (BLUP)* to predict breeding values and simultaneously to estimate heritability from pedigree data (Henderson 1973, 1976). This model has become known as the *animal model*. The use of breeding values will be discussed in more detail in Sect. 16.10. This section returns to heritability and its estimation using the animal model.

The additive genetic variance of an individual animal is modelled as a random effect in the (linear mixed) animal model. This means that this effect needs to be fitted for each individual (see Kruuk et al. 2014). The main methods for the animal model are the *Residual or Restricted Maximum Likelihood (REML)* and *Markov chain Monte Carlo (MCMC)* (Thompson et al. 2005; Wilson et al. 2010). Both methods are supported in the statistical software program R (R Core Team 2015), and will be presented in the next sections.

Assortative mating, selection and inbreeding do not result in biased estimates of variances in the animal model, unlike in parent–offspring regression (Kruuk 2004). Moreover, this model does not require breeding experiments, e.g. foster–parents, to estimate parental effects. The animal model "replaced" parent–offspring regression in ecological–evolutionary studies on wild populations by the end of the 1990s (see overview in Postma 2014).

The additive relationship (kinship) matrix is an important part of the animal model. The algorithms that were developed to calculate large kinship matrices (Henderson 1976; Quaas 1976) are used in studbook software to calculate inbreeding coefficients, founder representation and mean kinship values.[1] It is somewhat surprising that the animal model itself has not been explored as a tool to estimate heritability of wild traits in studbook populations, until recently (e.g. Pelletier et al. 2009).

16.9.1 REML

The R package *pedigreemm* (Bates and Vazquez 2014) is an extension to the linear mixed models in the R package *lme4* (Bates et al. 2014), and approximates the REML method in estimating variances. Gaussian (normal), binary (binomial) and count data (Poisson) are supported (Vazquez et al. 2010).

The phenotypic data in the animal model are linked to pedigree data through IDs that are unique for each individual (as in studbooks). The data field in the phenotypic data is generally named *id* or *animal* (which it is in this chapter). The animal model

[1]The modified algorithm by R. L. Quaas is used to compute inbreeding coefficients only.

Table 16.5 Heritability of parturition date in Nepalese red pandas in Europe and North America estimated with the REML animal model. Number of observed litters is 2,051, the number of breeding combinations (families) is 571

Model	V_A	V_{lat}	V_R	V_P	h^2	AIC	ΔAIC
Naive	62.2		151.6	213.8	0.29	16,473	115
Latitude	25.6	45.3	141.4	212.4	0.12	16,358	0

V_A additive genetic variance, V_{lat} variance due to latitude at location of birth, V_R residual variance, V_P total phenotypic variance, h^2 heritability, *AIC* Akaike Information Criterion, *ΔAIC* difference with model with lowest AIC

partitions the phenotypic trait of an individual in *fixed or predictable effects* and *random effects* (Kruuk et al. 2014). Sex and age are examples of fixed effects, while the effect of zoo locations on a trait could be considered as a random effect.

Correct interpretation of results of the animal model can be complex, as will be illustrated in this section. Therefore, it is recommend to start analysis with a "naive" (univariate) model in which only individuals are modelled as random effects. This model will therefore only estimate the additive genetic variance (V_A) and the residual variance (V_R), which together total the phenotypic variance (V_P). The residual variance (V_R) is the total of dominance variance (V_D), epistatic variance (V_I) and environmental variance (V_E).

Heritability in the animal model is estimated according to Eq. 16.6. See Vazquez et al. (2010) for details on the use of the pedigreemm software.

Table 16.5 presents results of analyses on the trait "parturition date" in Nepalese red pandas (restricted to locations in Europe and North America). Heritability estimated with the "naive" model is 0.29. This value is somewhere in the range of the (unadjusted) $h^2 = 0.36$ as estimated with mother–daughter regression (see Table 16.3).

Differences in results between the animal model and parent–offspring regression occur, especially when environmental effects are not modelled (de Villemereuil et al. 2013; Kruuk and Hadfield 2007). Spatial effects, such as sharing the same nest site or same home range as the parents, result in shared environmental conditions. This can result in non–genetic resemblance between relatives leading to overestimation of heritability (Stopher et al. 2012; van der Jeugd and McCleery 2002).

Nepalese red pandas adjust the birth season to the summer period in both northern and southern hemispheres (Table 3.8).Abiotic conditions related to latitude, e.g. photoperiod, may also adjust parturition date within hemispheres. This would mean that offspring which stay at the parental location or are moved to locations at a close distance share the same parturition date. This spatial environmental effect can be studied by modelling latitude as a random effect in the animal model.

When latitude is modelled, the additive genetic variation has been reduced by more than 50 %, resulting in a heritability of 0.12 (Table 16.5). This means that resemblance in this trait is partly caused by shared environmental conditions at a given latitude across generations (*transgenerational effects*).

Model selection can be based on the *Akaike Information Criterion (AIC)*. The "latitude" model has the lowest AIC value, and is considered as the "better" of the two models. The difference ΔAIC between each model is compared and evaluated. Models with a ΔAIC of about 9–11 have little support, and those models with $\Delta AIC > 20$ have no empirical support (Burnham et al. 2011). Since $\Delta AIC_{naive} = AIC_{naive} - AIC_{latitude} = 16,473 - 16,358 = 115$, the "naive" model is considered implausible.

16.9.2 MCMC

The R package *MCMCglmm* (Hadfield 2010) implements the MCMC method and follows Bayesian inference to estimate variances (and heritability). The strength of this method is its flexibility in analysing traits that do not follow the normal distribution (de Villemereuil et al. 2013; Morrisey et al. 2014).

Table 16.6 presents results of MCMC analysis for the same two models "naive" and "latitude" as used in the previous REML analyses. The additional model "permanent" includes the permanent environmental variance (V_{Eg}) and repeatability for the trait "parturition date".

A characteristic of Bayesian analysis is to use prior information. A "weak" prior is recommended for the animal model (Wilson et al. 2010), and is used for the analyses in this section. The number of iterations (runs) was set to 110,000 with the first 10,000 as *burn–in* period to ensure *convergence*. A thinning of 50 showed to be sufficient to reduce *autocorrelation* between runs (see Hadfield 2010; Wilson et al. 2011).

The pedigreemm package does not provide confidence intervals for heritability. In the Bayesian approach, a 95 % *credible interval* is provided. This interval is interpreted as the 95 % probability that a value is within the given range. For example, heritability in the "naive" model is between 0.21 and 0.37 (Table 16.6).

The results of the "naive" and "latitude" models do not differ from those using REML (Table 16.5). The *Deviance Information Criterion (DIC)* can, like AIC, be used to select the "better" model.

Table 16.6 Heritability of parturition date in Nepalese red pandas in Europe and North America, estimated with the MCMC animal model. Model parameters were 110,000 runs, a burn–in period of 10,000 and thinning of 50

Model	V_A	V_{lat}	V_{Eg}	V_R	V_P	h^2	r_P	DIC
Naive	62.8			152.0	214.8	0.29 [0.21, 0.37]		16,346
Latitude	25.7	46.0		141.7	213.4	0.12 [0.07, 0.17]		16,201
Permanent	13.5		48.0	135.7	197.2	0.07 [0.02, 0.12]	0.31 [0.26,0.36]	16,203

V_A additive genetic variance, V_{Eg} permanent environmental variance, V_{lat} variance due to latitude at location of birth, V_R residual variance, V_P total phenotypic variance, h^2 heritability (95 % credible interval in brackets), r_P repeatability (95 % credible interval in brackets) DIC Deviance Information Criterion

The third model "permanent" separates the between–individual variance into permanent environment and additive genetic variances (Wilson et al. 2010, 2011). Both heritability and repeatability can be estimated with this model. Heritability is considerably lower in the permanent model compared to the model with latitude as random effect i.e. h^2 is 0.07 and 0.12, respectively. The difference between DIC values of the "latitude" and "permanent" models, however, is low and there is no reason to select one of them as being "better".

The results of the "permanent" model are of the same order as estimated with parent–offspring (though significant at $\alpha = 0.1$), and adjusted single–parent regression analyses (see Table 16.3). This supports the findings of simulation experiments in which parent–offspring regression performs better than "naive" animal models (de Villemereuil et al. 2013).

16.9.2.1 Diagnostics of MCMC

It is important to check that the MCMC chains (iterations) converge properly and that successive iterations are not correlated. Poor convergence and autocorrelation lead to invalid posterior distributions of the fixed and random effects. The R package *coda* (Plummer et al. 2006) provides diagnostic tools to evaluate MCMC results (see Hadfield 2010; Wilson et al. 2010, 2011).

A graph with the density of the (posterior) distribution of heritability can already be useful for indicating potential problems. For example, a prior which is too weak may result in MCMC getting "stuck" in zero variances which will be visible when the density is concentrated near zero. Diagnosis of the different fixed and random effects is still recommended, even when the distribution of h^2 seems to be without problems.

Figure 16.3 shows the density distribution of heritability in the "permanent model". Since thinning is set to 50, the results of each 50th iteration after the burn-in of 10,000 is used. This means that density is effectively based on 2,000 iterations, which may explain why the distribution is not completely symmetrical. Nevertheless, values for the mean, median and mode heritability are very similar. In principle, the 95 % credible interval shows the accuracy of the "permanent model": heritability is expected to be between 0.02 and 0.12.

16.10 Monitoring Phenotypic Variation

Monitoring genetic variation in studbook populations through gene drop simulation, protein electrophoresis or microsatellites (DNA) provides information, whether actual or theoretical, on maintenance of variation in neutral genes. Arnold (1995) argued that maintenance of genetic variation in genes that are subject to selection is more important for the population than maintenance of neutral genes. Monitoring genetic variation in phenotypic traits which involve multiple genes is recommended as part of management of captive populations (e.g. Arnold 1995). This section

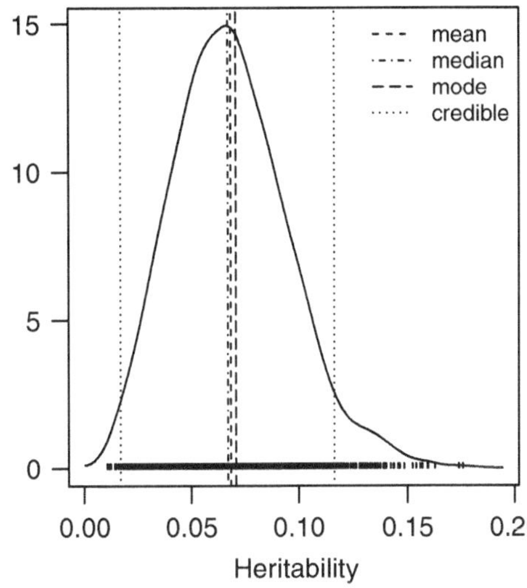

Fig. 16.3 Density distribution of heritability estimates in the "permanent" model. Model parameters were 110,000 runs, a burn–in period of 10,000 and thinning of 50. Density is plotted on the Y–axis

discusses techniques to monitor phenotypic traits in studbooks. Litter size in the European studbook population of African wild dogs (data period: 1986–2011) is used as an example.

It is logical that a monitoring scheme for pedigreed populations is based on (captive–born) generations (Chargé et al. 2014). The first step is to compare phenotypic variances per generation group. Box–and–whisker plots of trait values provide a graphic representation of Tukey five number summary (Tukey 1977) of phenotypic values within and between generation groups.

Figure 16.4a presents box–and–whisker plots of litter size among maternal generation groups in African wild dogs. Differences in the median and 25 % and 75 % quantiles (i.e. bottom and top of "the box") between generation groups can be observed in this figure. However, the major interest is to know whether (phenotypic) variances in litter size differ among generation groups. Levene's test for homogeneity of variances (Sokal and Rohlf 2012) can be used to test whether variances in litter size among generation groups are equal. This hypothesis is rejected ($F = 2.47$, $df = 6$, $p = 0.023$).

Although phenotypic variance of litter size differs among generations, no clear trend can be observed (Fig. 16.4a). This does not mean that no genetic trend exists. Genotypic change over time can be detected by comparing mean breeding values per year and/or generation (e.g. Chargé et al. 2014; Réale et al. 2003; Wilson et al. 2007). The animal model is the preferred method for estimating breeding values when (full) pedigree data are available, as discussed in Sect. 16.9.

The use of breeding values in ecological studies has been subject to critique. One of the criticisms is that overly simple models do not properly separate genetic and non–genetic phenotypic resemblance (Postma 2006), such as the effect of latitude on parturition date in red pandas. Non–genetic resemblance leads to overestimation of heritability and thus breeding values.

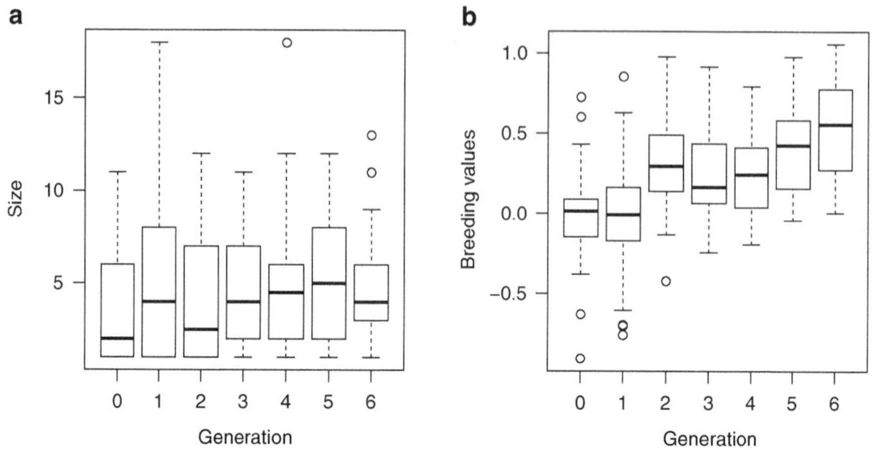

Fig. 16.4 (**a**) Box and whisker plots for litter size in maternal generation groups of African wild dogs (period 1986–2011). (**b**) Breeding values in maternal generation groups of African wild dogs (period 1986–2011)

Three models were analysed with the animal model to determine the appropriate model to use for estimating heritability and breeding values for litter size in African wild dogs. The "naive" model did not include extra random effects (other than "animal"); the "permanent" model estimates the effect of the permanent environmental variance (V_{Eg}) on heritability; the "extended" model also includes parental age (V_{age}) and litter year (V_{year}) as random effects. Sex and maternal generation were modelled as fixed effects in all three models.

The first model revealed that vague priors did not converge properly, i.e. "stuck at zero", when modelling permanent environmental variance. Therefore, a slightly informative prior was used, i.e. the observed phenotypic variance was equally distributed over the random (including residual) effects with a low degree of belief ($nu = 1$) (Chargé et al. 2014). Furthermore, the number of iterations was set to 350,000 with a burn–in of 50,000 and thinning of 50 in order to pass the various diagnostic tests for the extended model. These settings were applied to all three analyses.

Table 16.7 summarises the results of the three models. The heritability (h^2) of litter size in the African wild dog population is 0.13 in the "naive" model. However, this value is reduced to 0.06 when the permanent environmental variance (V_{Eg}) is modelled. Thus, some 50 % of the additive genetic variance (V_A) in the "naive" model should be attributed to the permanent environment. This non–genetic factor also explains a part of the residual variance (V_R) which reduced from 11.68 in the "naive" model to 10.22 in the "permanent" model. The random effects age of parents (V_{age}) and annual effects (V_{year}) do not have effect on heritability, but explain half the residual variance.

Table 16.7 Heritability of litter size in African wild dogs estimated with the MCMC animal model

Model	V_P	V_A	V_{Eg}	V_{age}	V_{year}	V_R	h^2
Naive	13.29	1.61				11.68	0.13 [0.05,0.20]
Permanent	13.23	0.79	2.23			10.22	0.06 [0.02,0.11]
Extended	13.30	0.74	2.20	4.46	1.21	4.69	0.06 [0.01,0.11]

V_P phenotypic variance, V_A additive genetic variance, V_{Eg} permanent environmental variance, V_{age} variance due to parental age, V_{year} variance due to year of litter, V_R residual variance, h^2 heritability [95 % credible interval in brackets]

The results in Table 16.7 re–emphasise the importance of using an appropriate model (Postma 2006): excluding the permanent environment results in overestimating heritability and breeding values.

The average breeding value for each individual was estimated from each run (after burn–in and thinning) in the extended model. Figure 16.4b shows box–and–whisker plots for these breeding values per generation. An upward trend in breeding values for litter size can be observed. Linear regression shows that this (linear) trend is statistically significant ($b = 0.085, se = 0.0088, p_t < 0.001$). These results indicate that (unconscious) selection for larger litter size has occurred in the captive population of African wild dogs.

It is not difficult to envision a scenario in the early history of the captive population when new packs started as a single pair, reproductive success was low and genetic management was not applied. Offspring of a few packs which successfully reared large litters, would quickly distribute the "large litter genes" through the population.

16.10.1 *Reliability of Breeding Values*

One needs to be cautious in drawing immediate conclusions without first assessing the *accuracy* of the estimated breeding values. The accuracy of estimated breeding values is defined as the correlation between variances in estimated and true breeding values (Postma 2006). The variance in true breeding values is the same as the additive genetic variance. The correlation (r) is estimated as:

$$r = \sqrt{\frac{V_{EBV}}{V_A}} \tag{16.10}$$

where V_{EBV} and V_A are the variances in estimated and true breeding values, respectively (Postma 2006).

The values for V_{EBV} and V_A in the extended model are 0.12 and 0.74, respectively. This means that the accuracy of EBV is $\sqrt{(0.12/0.74)} = 0.40$. The interpretation of accuracy is easier when expressed as *reliability*, which is the squared correlation r^2. This value equals the proportion of additive genetic variance that is accounted for by EBVs (V_{EBV}/V_A) (Postma 2006).

Reliability of estimated breeding values for litter size in the African wild dog population is $0.12/0.74 = 0.16$. Given this low value one needs to be careful in drawing conclusions from estimated breeding values regarding a selection gradient. In other words, unconscious selection for large litter size may not have taken place.

Reliability is high when heritability is high,[2] and, furthermore, increases when more phenotypic information (including on close relatives) on individuals is available. Low heritability in litter size and the fact that nearly 40 % of the reproducing individuals produced a single litter explains the low reliability of estimated breeding values for this trait in the African wild dog population.

16.11 Remarks

Heritabilities of life history traits are, across species, the lowest compared to physiological, behavioural and morphological traits (Postma 2014). The results of analyses in this chapter involving the traits "fitness", "litter size" and "parturition date" are in line with these findings. The low heritabilities may make life history traits less attractive to apply in standard monitoring schemes of phenotypic variance than, for example, morphological traits with mean heritability of 0.56 ± 0.035 (Postma 2014). However, the disadvantage of low heritability needs to be balanced with the fact that pedigree *and* phenotypic data can be directly accessed from computerised studbooks. Monitoring schemes involving morphological traits require extra data collection per studbook and most likely lack historical data, e.g. on wild–born and early generations.

Studbooks on threatened species are small compared to pedigrees maintained for livestock. However, various studbooks started some 25–30 years ago, when regional zoo associations initiated cooperative management programmes, and are gradually becoming large enough for quantitative genetic analysis. Maybe the biggest challenge is modelling the right effects on phenotypic variance, as zoo species are distributed all over the globe with different environments, while at the same time in–depth knowledge of life history for many species is still a mystery.

[2]The minimum reliability equals heritability when only a single record for an individual is available.

References

Åkesson M, Bensch S, Hasselquist D, Tarka M, Hansson B (2008) Estimating heritabilities and genetic correlations: comparing the 'animal model' with parent–offspring regression using data from a natural population. PLoS ONE 3(3):e1739

Arnold SJ (1994) Multivariate inheritance and evolution: a review of concepts. In: Boake CR (ed) Quantitative genetic studies of behavioral evolution. University of Chicago Press, Chicago, pp 17–48

Arnold SJ (1995) Monitoring quantitative genetic variation and evolution in captive populations. In: Ballou et al. (1995), pp 295–317

Ballou JD, Gilpin M, Foose TJ (eds) (1995) Population management for survival and recovery: analytical methods and strategies in small population conservation. Methods and cases in conservation science series. Columbia University Press, New York

Bates D, Vazquez AI (2014) pedigreemm: pedigree-based mixed-effects models. http://CRAN.R-project.org/package=pedigreemm, R package version 0.3-3

Bates D, Maechler M, Bolker B, Walker S (2014) lme4: linear mixed–effects models using Eigen and S4. http://CRAN.R-project.org/package=lme4, R package version 1.1-7

Box GEP, Cox DR (1964) An analysis of transformations. J R Stat Soc Ser B (Methodol) 26(2):211–252

Burnham KP, Anderson DR, Huyvaert KP (2011) AIC model selection and multimodel inference in behavioral ecology: some background, observations, and comparisons. Behav Ecol Sociobiol 65(1):23–35

Chargé R, Sorci G, Saint Jalme M, Lesobre L, Hingrat Y, Lacroix F, Teplitsky C (2014) Does recognized genetic management in supportive breeding prevent genetic changes in life–history traits? Evol Appl 7(5):521–532

Charmantier A, Garant D, Kruuk LEB (eds) (2014) Quantitative genetics in the wild. Oxford University Press, Oxford

Crawley MJ (2007) The R book. Wiley, Chichester

Crow JF, Kimura M (1970) An introduction to population genetics theory. Harper and Row, New York

de Villemereuil P, Gimenez O, Doligez B (2013) Comparing parent—offspring regression with frequentist and Bayesian animal models to estimate heritability in wild populations: a simulation study for Gaussian and binary traits. Methods Ecol Evol 4(3):260–275

Dohm MR (2002) Repeatability estimates do not always set an upper limit to heritability. Funct Ecol 16(2):273–280

Falconer DS (1960) Introduction to quantitative genetics, 1st edn. Longman, London

Fisher RA (1918) The correlation relatives on the supposition of Mendelian inheritance. Trans R Soc Edinb 52:399–433

Fox J, Weisberg S (2011) An R companion to applied regression, 2nd edn. Sage, Thousand Oaks

Frankham R (1999) Quantitative genetics in conservation biology. Genet Res 74:237–244

Frankham R (2008) Genetic adaptation to captivity in species conservation programs. Mol Ecol 17(1):325–333

Hadfield JD (2010) MCMC methods for multi-response generalized linear mixed models: the MCMCglmm R package. J Stat Softw 33(2):1–22

Henderson CR (1973) Sire evaluation and genetic trends. J Anim Sci 1973:10–41

Henderson CR (1976) A simple method for computing the inverse of a numerator relationship matrix used in prediction of breeding values. Biometrics 32:69–83

Houle D (1992) Comparing evolvability and variability of quantitative traits. Genetics 130(1):195–204

Hurlbert SH (1984) Pseudoreplication and the design of ecological field experiments. Ecol Monogr 54(2):187–211

Ibáñez B, Cervantes I, Gutiérrez JP, Goyache F, Moreno E (2014) Estimates of direct and indirect effects for early juvenile survival in captive populations maintained for conservation purposes: the case of Cuvier's gazelle. Ecol Evol 4(21):4117–4129
Keller LF, Grant PR, Grant R, Petren K (2001) Heritability of morphological traits in Darwin's finches: misidentified paternity and maternal effects. Heredity 87(3):325–336
Kempthorne O, Tandon OB (1953) The estimation of heritability by regression of offspring on parent. Biometrics 9(1):90–100
Kimura M (1968) Evolutionary rate at the molecular level. Nature 217:624–626
Kruuk LEB (2004) Estimating genetic parameters in natural populations using the 'animal model'. Philos Trans R Soc B 359:873–890
Kruuk LEB, Hadfield JD (2007) How to separate genetic and environmental causes of similarity between relatives. J Evol Biol 20(5):1890–1903
Kruuk LEB, Clutton–Brock TH, Slate J, Pemberton JM, Brotherstone S, Guinness FE (2000) Heritability of fitness in a wild mammal population. Proc Natl Acad Sci USA 97(2):698–703
Kruuk LEB, Charmantier A, Garant D (2014) The study of quantitative genetics in wild populations. In: Charmantier et al. (2014), pp 1–15
Lande R (1995) Breeding plans for small populations based on the dynamics of quantitative genetic variance. In: Ballou et al. (1995), pp 318–340
Lessells CM, Boag PT (1987) Unrepeatable repeatabilities: a common mistake. Auk 104(1):116–121
McAdam A, Garant D, Wilson AJ (2014) The effect of others' genes: maternal and other indirect effects. In: Charmantier et al. (2014), pp 84–103
McComb B, Zuckerberg B, Vesely D, Jordan C (2010) Monitoring animal populations and their habitats. CRC Press, Boca Raton
McDougall PT, Réale D, Sol D, Reader SM (2006) Wildlife conservation and animal temperament: causes and consequences of evolutionary change for captive, reintroduced and wild populations. Anim Conserv 9(1):39–48
Morrisey MB, de Villemereuil P, Doligez B, Gimenez O (2014) Bayesian approaches to the quantitative genetic analysis of wild populations. In: Charmantier et al. (2014), pp 228–253
Mrode RA (2000) Linear models for the prediction of animal breeding values. CAB International, Wallingford
Nakagawa S, Schielzeth H (2010) Repeatability for Gaussian and non–Gaussian data: a practical guide for biologists. Biol Rev 85(4):935–956
Pelletier F, Réale D, Watters J, Boakes EH, Garant D (2009) Value of captive populations for quantitative genetics research. Trends Ecol Evol 24(5):263–270
Plummer M, Best N, Cowles K, Vines K (2006) Coda: convergence diagnosis and output analysis for mcmc. R News 6(1):7–11. http://CRAN.R-project.org/doc/Rnews/
Postma E (2006) Implications of the difference between true and predicted breeding values for the study of natural selection and micro–evolution. J Evol Biol 19(2):309–320
Postma E (2014) Four decades of estimating heritabilities in wild vertebrate populations: improved methods, more data, better estimates? In: Charmantier et al. (2014), pp 16–33
Quaas RL (1976) Computing the diagonal elements and inverse of a large numerator relationship matrix. Biometrics 32:949–953
R Core Team (2015) R: a language and environment for statistical computing. R Foundation for Statistical Computing, Vienna. https://www.R-project.org/
Réale D, McAdam AG, Boutin S, Berteaux D (2003) Genetic and plastic responses of a northern mammal to climate change. Proc R Soc Lond B Biol Sci 270(1515):591–596
Reeve ECR (1955) The variance of the genetic correlation coefficient. Biometrics 11(3):357–374
Ricklefs RE, Cadena CD (2008) Heritability of longevity in captive populations of nondomesticated mammals and birds. J Geront Ser A Biol Sci Med Sci 63(5):435–446
Schielzeth H, Nakagawa S (2013) rptR: repeatability for Gaussian and non-Gaussian data. http://R-Forge.R-project.org/projects/rptr/, R package version 0.6.405/r52
Sokal RR, Rohlf FJ (2012) Biometry, 4th edn. W.H. Freeman and Company, New York

Stopher KV, Walling CA, Morris A, Guinness FE, Clutton-Brock THC, Pemberton JM, Nussey DH (2012) Shared spatial effects on quantitative genetic parameters: accounting for spatial autocorrelation and home range overlap reduces estimates of heritability in wild red deer. Evolution 66(8):2411–2426

Thompson R, Brotherstone S, White IMS (2005) Estimation of quantitative genetic parameters. Philos Trans R Soc Lond 360:1469–1477each sex. This method requires

Tukey JW (1977) Exploratory data analysis. Behavioural science: quantitative methods. Addison–Wesley, Reading

van der Jeugd HP, McCleery R (2002) Effects of spatial autocorrelation, natal philopatry and phenotypic plasticity on the heritability of laying date. J Evol Biol 15(3):380–387

Vazquez AI, Bates DM, Rosa GJM, Gianola D, Weigel KA (2010) Technical note: an R package for fitting generalized linear mixed models in animal breeding. J Anim Sci 88(2):497–504

Visscher PM, Hill WG, Wray NR (2008) Heritability in the genomics era – concepts and misconceptions. Nat Rev Genet 5(4):255–266

Vittinghoff E, Glidden DV, Shiboski SC, McCulloch CE (2005) Regression methods in biostatistics. Statistics for biology and health. Springer, New York

Weis AE (2005) Direct and indirect assortative mating: a multivariate approach to plant flowering schedules. J Evol Biol 18(3):536–546

Wilson AJ, Pemberton JM, Pilkington JG, Clutton–Brock TH, Coltman DW, Kruuk LEB (2007) Quantitative genetics of growth and cryptic evolution of body size in an island population. Evol Ecol 21(3):337–356

Wilson AJ, Réale D, Clements MN, Morrissey MM, Postma E, Walling CA, Kruuk LEB, Nussey DH (2010) An ecologist's guide to the animal model. J Anim Ecol 79(1):13–26

Wilson AJ, Réale D, Clements MN, Morrissey MM, Postma E, Walling CA, Kruuk LEB, Nussey DH (2011) An ecologist's guide to the animal model – erratum. J Anim Ecol 80:1109

Wolak ME, Fairbairn DJ, Paulsen YR (2012) Guidelines for estimating repeatability. Methods Ecol Evol 3(1):129–137

Wright S (1921a) Systems of mating. III. Assortative mating based on somatic resemblance. Genetics 6(2):144–161

Wright S (1921b) Systems of mating. IV. The effects of selection. Genetics 6(2):162–166

Part IV
Conservation

Chapter 17
Conservation

Abstract Natural history data are important in conservation of endangered species. The public databases AnAge (animal longevity) and COMADRE (matrix population models) were analysed for the origin of data sources. The results show that information on natural history elements, including longevity, for most wild species are actually derived from captive populations. The availability of life tables based on wild populations is still limited. The analyses show that the average time to obtain life history data on wild populations is 16 years. Therefore, captive populations are important sources for base–line data on endangered wild species. The differences between captive and wild environments and potential effects on natural history elements are discussed. A Population Viability Analysis (PVA) that was conducted on wild lowland bongo (antelope) in Dzanga Sangha Protected Areas, Central African Republic serves as an example to illustrate applicability of captive data to field conservation decisions. Sensitivity analysis to evaluate assumptions resulted in more realistic scenarios to assess population viability. The chapter ends with a section justifying the (proper) use of captive data in conservation of wild populations.

17.1 Wild Data

The primary contribution of studbooks to conservation of endangered species is providing data for scientific management of captive and semi–wild populations, whether maintained for educational purposes, research or reintroduction. The same data can also be valuable resources for management of wild populations, especially when natural history data on these populations are incomplete.

Chapter 3 presented analyses to "extract" natural history elements, such as (reproductive) lifespan, seasonality and litter size, from studbook data, that could potentially be used in management of wild populations. Since captive and wild environments do differ, it is a valid question to ask whether such an approach is justified.

Field data are missing or incomplete for many species, depending on the taxonomic group. This makes it difficult to provide a straightforward answer to the above question. The *AnAge* database contains data on longevity and other natural history elements on more than 4,000 vertebrate species (de Magalhães and Costa

F.P.G. Princée, *Exploring Studbooks for Wildlife Management and Conservation*, Topics in Biodiversity and Conservation 17, DOI 10.1007/978-3-319-50032-4_17

Table 17.1 Origin of data sources for vertebrate species in the AnAge (Tacutu et al. 2013) and COMADRE (Salguero–Gómez et al. 2016) databases. The data are grouped per animal class

Class		Anage database[a]		COMADRE database	
Scientific name	English name	Wild	Captive	Wild	Captive
Cephalaspidomorphi	Lampreys	15	–	–	–
Chondrichthyes	Sharks, skates and rays	112	1	–	–
Osteichthyes[b]	Bony fish	759	34	49	6
Amphibia	Amphibians	28	125	9	–
Reptilia	Reptiles	15	490	24	–
Aves	Birds	720	335	60	–
Mammalia	Mammals	80	922	92	3

[a] Data sources refer to those used for longevity records
[b] Comprises subclasses Actinopterygii and Sarcopterygii

2009; Tacutu et al. 2013). Data quality and sample size are assessed as well as the origin of specimens in longevity records (i.e. captive, wild or unknown). Table 17.1 presents an overview by vertebrate class of the origin of longevity records. Records based on specimens of unknown origin are not included in this overview.

Table 17.1 shows the differences between vertebrate classes in the contribution of field (wild) and captive data to longevity records. Field data are the primary source (i.e. 96 %) in the three "fish" classes. A striking difference is the contribution of the field data for birds (68 %), especially through banding studies, compared to amphibians, reptiles and mammals i.e. 18, 3 and 8 %, respectively.

Longevity can be used as an indicator of viability in conservation of endangered species. The detailed information embedded in age–specific life tables, among other data such as age distribution, are preferred for assessing population viability. However, the number of available data-sets from field studies is low compared to the number of longevity records (and the number of endangered species). The *COMADRE Animal Matrix Database* includes *Matrix Population Models (MPMs)* on some 267 animal taxa collated from different sources (Salguero–Gómez et al. 2016). Table 17.1 shows that the life tables in this database are almost all based on field data.

Life tables are preferably constructed from longitudinal studies (Nussey et al. 2008). It is very likely that life table data on most wild populations are not available. For example, field data on only 27 carnivore taxa were available for a study on population dynamics (van de Kerk et al. 2013). This is less than 10 % of the 288 extant carnivore species (IUCN 2016). Some 25 % of the taxa ($n = 7$) in these population studies were listed in one of the threatened categories of the IUCN Red List.

Long–term studies are not always practical or feasible when species are threatened with extinction, as mentioned previously. For example, the mean duration of field studies in the COMADRE database is 16 years. Management measures that require results from long–term studies may simply come too late. Life tables based on captive populations, i.e. from international and regional (zoo) studbooks and/or

ISIS data, need to be considered as alternative data sources when field data are not available. The next sections will discuss the extent to which captive and wild data are likely to differ, and how to handle these differences.

17.2 Captive and Wild Environments

Captive and wild environments obviously differ. Animals in captivity are expected, in the absence of predators, with stable food and water supply, and veterinary care, to live longer than their wild conspecifics. A comparative study on longevity in species for which both captive and wild sources were available supports this expectation (de Magalhães et al. 2007). Studies in which captive and wild populations were compared also showed higher (adult) survivorship in captivity (Lynch et al. 2010; Ricklefs and Scheuerlein 2001).

It is important to evaluate captive data per species (or studbook population). Incomplete information on natural history and/or limited experience with holding a specific species in captivity can result in suboptimal husbandry. This situation tended to occur in the period of time before intensive cooperation between zoological institutions was initiated. For example, a comparative study in Sumatran orang utans (*Pongo abelii*) showed that survivorship of captive orang utans in the period 1946–1965 was lower than in the wild. However, no differences were found when more recent studbook data (i.e. 1986–2005) were used (Wich et al. 2009).

Management measures can have an impact on life history of zoo populations. For example, cooperative management programmes in which contraception is applied will (temporarily) result in reduced fecundity rates. This means that an appropriate date span needs to be selected whenever results of studbook analyses are applied to wild populations.

The fact that natural history elements can differ between captive and wild populations does not mean that these data cannot be used. Natural history data can also differ between wild populations, e.g. survivorship in wild chimpanzee (*Pan troglodytes*) populations (Hill et al. 2001). Moreover, wild populations of endangered species can be compromised and not necessarily represent the original "natural" state. For example, severe poaching in a population of African elephants (*Loxodonta africana*) resulted in a shift in mean age of females at first reproduction from 16 years to 11.3 years (Owens and Owens 2009). Small wild populations can be subject to inbreeding and inbreeding depression (e.g. Keller and Waller 2002). Therefore, the issue is not "whether" but "how" captive data should be used in conservation management.

The next section uses a population viability study on lowland bongo (*Tragelaphus eurycerus eurycerus*) in Dzanga Sangha Protected Areas, Central African Republic (Princée 2011), to illustrate the combined use of captive and wild data.

17.3 Population Viability Analysis

Population Viability Analysis (PVA) is a model used to assess risks that populations go extinct during a specified period of time. The model generally includes stochastic processes that have impact on populations: demographic, genetic, and environmental processes and catastrophes. The interactions between these processes are complex and computer software is required to apply a PVA (see Brook et al. 2000). The computer program *VORTEX* which is used by the IUCN/SSC Conservation Breeding Specialist Group implements the PVA model (Lacy 2000; Lacy et al. 2014). VORTEX data and descriptions in this section refer to the older VORTEX version 9.99b (Lacy et al. 2009) that was used in the bongo PVA (Princée 2011).

Although the bongo (*Tragelaphus eurycerus*) is a large antelope, with males weighing from 250 to 400 kg (Kingdon 2015), its secretive lifestyle means that many of the natural history elements, such as gestation length, inter–birth interval and life history, are known only from information on the mountain bongo (*Tragelaphus eurycerus isaaci*) subspecies in captivity (Bosley 2010).

Data from daily sightings of lowland bongo in Dzanga Baie (Turkalo and Klaus-Hügi 1999; Turkalo unpublished data) and life history data from the mountain bongo studbook were compared and combined to create a base–line data set for the PVA.

17.3.1 Studbook Data

Computerised data of the International studbook on Eastern/Mountain Bongo, updated until 31 December 2010, contained 2,305 records (Bosley 2011) and was used for the bongo PVA. Analyses included only individuals that were born in the period 1970–2010 or living on 1 January 1970. Table 17.2 presents the use of studbook life history data in VORTEX. These data are not directly entered as life tables, but need some transformation (and decision–making).

17.3.1.1 Reproduction

The maximum age of reproduction can be based on the oldest fecundity class with a sufficient sample size, or on a percentile of observed ages at birth. The VORTEX model assumes that individuals die once reaching the post–reproductive age, i.e. class 15+.

Fecundity is not entered as age specific fecundity rates (m_x) in VORTEX but as a combination of the maximum progeny per female per year, i.e. litter size and number of litters per year, and the proportion of adult females breeding per year.

Studies on lowland bongo in Dzanga Sangha PA and the adjacent Nouabali-Ndoki National Park (Republic of Congo) suggest that this species mates during

Table 17.2 Studbook life history data used the basic scenario ($A1$) of the lowland bongo PVA. Data source: Princée (2011). The data entry sequence of VORTEX 9.99b is followed. See text for details

VORTEX data	Value	Remarks
Reproductive system		
Age at first offspring females	3 years	Median age at first reproduction
Age at first offspring males	4 years	
Maximum age of reproduction	14 years	Based on 97.5 % percentile of age at birth
Maximum number of litters per year	1	Indicated by a single breeding season in the wild
Maximum number of progeny per year	1	Studbook: monotocous
Sex–ratio at birth in % males	45	Significantly different from an equal sex-ratio ($p < 0.001$)
Reproductive rates		
% adult female breeding	0.55	Census annual births/adult female
SD in % breeding due to EV	0.07	Standard deviation annual births/adult female rate
Distribution of litters: litter 1	100 %	
Mortality rates		
Mortality of females as %		
Mortality from age 0 to 1	21.6	Studbook mortality rates
Mortality from age 1 to 2	6.1	
Mortality from age 2 to 3	6.3	
Annual mortality after 3	7.6	Geometric mean classes 3–14
Mortality of males as %		
Mortality from age 0 to 1	27.3	Studbook mortality rates
Mortality from age 1 to 2	11.1	
Mortality from age 2 to 3	9.50	
Mortality from age 3 to 4	10.5	
Annual mortality after 4	13.3	Geometric mean classes 4–14

the major dry season (December–February) (Turkalo and Klaus-Hügi 1999). The calving season is during the major rainy season (September–November) (Elkan et al. 2009). Thus females in the natural habitat are expected to give birth once a year.

The bongo is clearly a monotocous species: only 10 twins and 1 triplet out of 1,945 births have been observed in the studbook population. It is assumed that the same applies to the wild population. This means that with a single birth season, the maximum number of progeny per year is 1.

Sex-ratio at birth ($M/(M + F)$) is 0.45 ($n = 2,166$) which differs significantly from an equal sex-ratio ($\chi^2 = 19.97, p < 0.001$) (see Table 17.2).

The percentage of adult females breeding per year can be derived from the mean inter–birth interval (Miller and Lacy 2005). For example, the mean inter–birth interval in the bongo studbook is 517 days ($\approx 16\frac{1}{2}$ months). This means that only

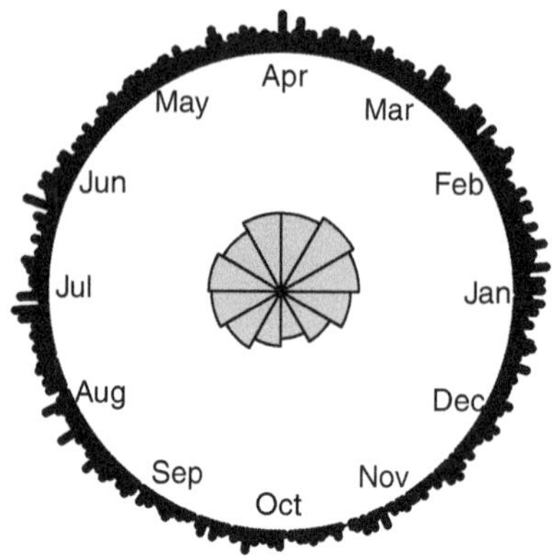

Fig. 17.1 Season of birth in mountain bongo in northern hemisphere zoos. Stacked points refer to births per day of year; the area sectors in the rose diagram represents relative frequencies of births per month

a proportion of $12/16.5 = 0.72$ (72 %) of the adult females breed per year. This percentage, however, is likely to be an overestimate as it refers to females that have reproduced at least twice! Adult females which have not reproduced are clearly not included in inter–birth analyses.

A better approach is to estimate the annual percentage of breeding females from census data (see Chap. 5). Since bongo in (northern hemisphere) zoos breed all year round (see Fig. 17.1), the census date is set to 1 January. The number of adult females, i.e. the reproductive age group from 3 to 14 years, are "counted" at each census date. Since bongo are monotocous, the rate "births per year/adult females" directly refers to the proportion of females that are breeding in a given year.

The arithmetic mean, weighted for number of adult females, over the annual rates in the period 1970–2010 is 0.55 ($SD = 0.07$). This means that on average 55 % of the adult females are expected to reproduce in a given year (Table 17.2). This value is close to the breeding ratio of 58 % as estimated from data from the natural clearing Dzanga Baie (Turkalo and Klaus-Hügi 1999).

17.3.1.2 Mortality

Age specific mortality rates are entered for the pre–reproductive age classes in VORTEX (see Table 17.2). A constant mortality rate (q_{adult}) is assumed in the reproductive age classes (same for both sexes). Mortality in the last reproductive age class is 1.0 (VORTEX models reproduction before death).

The constant q_{adult} in the bongo PVA was estimated as the geometric mean of mortality rates in the reproductive age classes, weighted against sizes of age classes (Princée 2011). An alternative method to calculate this constant is described in Chap. 19.

17.3.1.3 Inbreeding Depression

VORTEX can model inbreeding depression through lethal equivalents (see Chap. 13). Lethal equivalents for 6–month survival in the bongo studbook population was estimated as 0.88 with maximum likelihood methods (Princée 2011). This value is used in the basic scenario.

17.3.2 Sensitivity Analysis

The VORTEX model includes a large range of parameters that can impact population viability. Since the focus of this section is on parameters obtained from studbooks, other simulation parameters are not considered. The number of VORTEX runs in the bongo PVA was 100 over 100 years; initial population size was assumed to be 300 animals with a stable age distribution; the carrying capacity was set to the arbitrary number of 750 in order to illustrate exponential growth.

Figure 17.2 presents the results of the basic bongo scenario *A*1. The population grows within 15 years to the (arbitrary) carrying capacity. This fast growth is not that surprising given that the life history data are based on a studbook population which grew from 29 to 630 animals in the period 1970–2010.

Daily sightings of lowland bongo in Dzanga Baie have been registered since 1991 (Turkalo and Klaus-Hügi 1999; Turkalo unpublished data). Figure 17.3 shows the largest observed daily count for each year. These data suggest that the population started to increase in 2008 (the year after trophy hunting was stopped in an adjacent forest reserve). However, it not clear whether trends in the baie can be fully extrapolated to the entire population (Princée 2011).

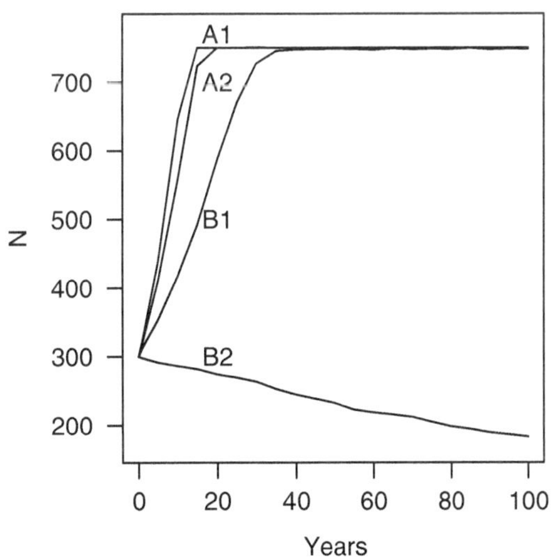

Fig. 17.2 VORTEX simulation results for mean census size under bongo PVA scenarios *A*1, *A*2, *B*1 and *B*2. See text for further details

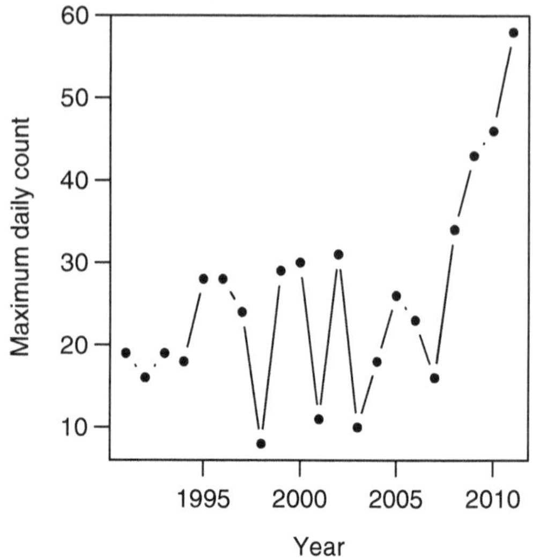

Fig. 17.3 Maximum observed daily counts of bongo per year in Dzanga Baie between 1991 and 2011 (Data: Andrea Turkalo)

A scenario that projects growth seems to be a realistic model. The main question is to determine how fast the population will grow. An important aspect of population modelling is sensitivity analysis to assess the impact of individual parameters.

The maximum age at reproduction in scenario $A2$ is changed from 14 to 12 years (which is the 95 percentile in male bongo). The simulation results are presented in Fig. 17.2. The population is still fast growing, but reaches carrying capacity some 5 years later than under scenario $A1$. Since maximum lifespan in captivity is longer than in the wild population (de Magalhães et al. 2007), reducing the maximum reproductive lifespan seems to be justified.

Juvenile (first year) mortality rates of 0.22 and 0.27 in the captive bongo population are in the range of observed rates in captive hoofstock (Kohler et al. 2006). The rates for captive bongo are considerably lower than those estimated for wild populations of hoofstock. For example, juvenile mortality rates estimated for African ungulates in Kruger National Park (KNP), South Africa are in the order of 0.4–0.6 (Owen–Smith and Mason 2005). The leopard (*Panthera pardus*) is assumed to be the main predator on calves in Dzanga Sangha PA (Turkalo and Klaus-Hügi 1999). Therefore, juvenile mortality that better reflects natural rates was deemed necessary in the bongo PVA (Princée 2011).

A general problem in field studies on life history is to assess neonatal and early mortality. For example, juveniles in the KNP study were already 6–12 months old when the annual surveys took place (Owen–Smith and Mason 2005). Lowland bongo have a hiding strategy for calves, which thus may not be observed until a few weeks old (Elkan 2003). However, most first year mortality in captive bongo occurs in the first week of life i.e. $q_{week1} = 0.18$. This mortality needs to be included in any juvenile mortality rate.

Differences between calf/adult female ratios since the birth (= major rainy) season could not be applied, as it was unknown whether increase of females without calves in the minor dry season, when the number of sightings also increases, referred to non–reproductive females or females which lost their calves (see Princée 2011). In the end, juvenile mortality was increased to 0.35 and 0.5 in scenarios *B*1 and *B*2, respectively.

Figure 17.2 presents the simulation results of all scenarios. The high juvenile mortality in scenario *B*2 results in a rapid decline of the population. Since annual sightings were increasing, a population increase seems to be more likely (see Fig. 17.3). Scenario *B*1 seems better to represent life history of the wild bongo population than the other three scenarios.

This example shows the importance of sensitivity analysis in adjusting parameters obtained from other populations (whether captive or wild) in order better to reflect local population dynamics. Although no detailed (wild) life tables were available on any lowland bongo population, the combined studbook data from mountain bongo and Dzanga Baie observations on lowland bongo provided a more realistic scenario.

Life tables are not the only factors that impact on the viability of bongo. However, most of the other data, such as initial population size (and carrying capacity) and poaching rates, need to be obtained from the wild.

17.4 Justification

Studbook data are often the only available life history data for conservation of endangered species. Ignoring these data because they can differ from an unknown wild population seems not to be justified, but nor is blindly applying them. It is important to understand the effect of the captive environment (including management measures) on natural history elements in order to make appropriate adjustments before applying them to wild populations.

The captive environment is not a single environment but a collection of zoological institutions in different climate zones and at different latitudes with different husbandry regimes. Although endangered species are certainly not part of experiments, the diversity in captive environments could be exploited to reveal features that otherwise could not be observed in the wild. For example, seasonality analysis of studbook data shows that mountain bongo have the potential to breed all year round (see Fig. 17.1). This suggests that the mating and birth season in the wild is largely determined by environmental factors. This would mean that bongo can adjust to shifts in seasons, e.g. a delayed dry (=mating) season.

Studbooks that cover zoological institutions in different climate zones provide data to study effects of temperature and/or precipitation on life history. For example, neonatal survival in captive red pandas is compromised in warmer climates (Princée and Glatston 2016). Results of such studies are not only important for husbandry, but can also serve as indicators of a species' vulnerability to climate change in the natural habitat.

Although there is sufficient empirical evidence that inbreeding depression occurs in (small) wild populations (Allendorf et al. 2013; Frankham et al. 2010) studies on captive populations will continue to contribute to the knowledge on this subject. This could, for example, include methods to detect inbreeding effects that manifest during an individual's lifetime (see Chap. 13).

In–depth understanding of the effects of the captive environment on natural history elements is not only important for management of captive populations, but also important to determine proper use of these data in wildlife conservation. Studies that partition the captive environment into specific features (e.g. location, nutrition, group size) and that assess their individual effects would be required. This may mean more exploring of studbook data and available tools in modern statistical software.

References

Allendorf FW, Luikart G, Aitken SN (2013) Conservation and the genetics of populations, 2nd edn. Wiley-Blackwell, Chichester

Bosley LF (2010) International studbook for Eastern/Mountain Bongo (*Tragelaphus eurycerus isaaci*) – Year 2009 Historical Edition, vol XXIV. Logsden, OR

Bosley LF (2011) International studbook for Eastern/Mountain Bongo (*Tragelaphus eurycerus isaaci*) dataset. Technical report, Logsden, OR

Brook BW, O'Grady JJ, Chapman AP, Burgman MA, Akçakaya HR, Frankham R (2000) Predictive accuracy of population viability analysis in conservation biology. Nature 404:385–387

de Magalhães JP, Costa J (2009) A database of vertebrate longevity records and their relation to other life–history traits. J Evol Biol 22(8):1770–1774

de Magalhães JP, Costa J, Church GM (2007) An analysis of the relationship between metabolism, developmental schedules, and longevity using phylogenetic independent contrasts. J Gerontol Ser A: Biol Sci Med Sci 62(2):149–160

Elkan PW (2003) Ecology and conservation of bongo antelope (*Tragelaphus eurycerus*) in lowland forest, Northern Republic of Congo. Ph.D. thesis, University of Minnesota, Minneapolis

Elkan PW, Parnell R, Smith JLD (2009) A die-off of large ungulates following a *Stomoxys* biting fly out-break in lowland forest, Northern Republic of Congo. Afr J Ecol 47(4):528–536

Frankham R, Ballou JD, Briscoe DA (2010) Introduction to conservation genetics, 2nd edn. Cambridge University Press, Cambridge

Hill K, Boesch C, Goodall J, Pusey A, Williams J, Wrangham R (2001) Mortality rates among wild chimpanzees. J Hum Evol 40(5):437–450

IUCN (2016) The IUCN red list of threatened species, version 2016-1. http://www.iucnredlist.org

Keller LF, Waller DM (2002) Inbreeding effects in wild populations. Trends Ecol Evol 17(5):230–241

Kingdon J (2015) The Kingdon field guide to African mammals, 2nd edn. Bloomsbury Publishing, London

Kohler IV, Preston SH, Lackey LB (2006) Comparative mortality levels among selected species of captive animals. Demogr Res 15:413–434

Lacy RC (2000) Structure of the VORTEX simulation model for population viability analysis. Ecol Bull 48:191–203

Lacy RC, Borbat M, Pollak JP (2009) VORTEX: a stochastic simulation of the extinction process, version 9.96. Technical report, Chicago Zoological Society, Brookfield

Lacy RC, Miller PS, Traylor–Holzer K (2014) VORTEX 10 User's manual. IUCN/ SSC Conservation Breeding Specialist Group, Apple Valley

Lynch HJ, Zeigler S, Wells L, Ballou JD, Fagan WF (2010) Survivorship patterns in captive mammalian populations: implications for estimating population growth rates. Ecol Appl 20(8):2334–2345

Miller PS, Lacy RC (2005) VORTEX: a stochastic simulation of the extinction process. Version 9.50 user's manual. IUCN/SSC Conservation Breeding Specialist Group, Apple Valley

Nussey DH, Coulson T, Festa–Bianchet M, Gaillard JM (2008) Measuring senescence in wild animal populations: towards a longitudinal approach. Funct Ecol 22(3):393–406

Owen–Smith N, Mason DR (2005) Comparative changes in adult vs. juvenile survival affecting population trends of African ungulates. J Anim Ecol 74(4):762–773

Owens MJ, Owens D (2009) Early age reproduction in female savanna elephants (*Loxodonta africana*) after severe poaching. Afr J Ecol 47:214–222

Princée FPG (2011) Population viability assessment western/lowland bongo in Dzanga–Sangha Protected Areas. Technical report, WWF-CAR, Bayanga, Central African Republic

Princée FPG, Glatston AR (2016) Influence of climate on the survivorship of neonatal red pandas in captivity. Zoo Biol 35(2):104–110

Ricklefs R, Scheuerlein A (2001) Comparison of aging-related mortality among birds and mammals. Exp Gerontol 36(4–6):845–857

Salguero–Gómez R, Jones OR, Archer CR, Bein C, de Buhr H, Farack C, Gottschalk F, Hartmann A, Henning A, Hoppe G, Römer G, Ruoff T, Sommer V, Wille J, Voigt J, Zeh S, Vieregg D, Buckley YM, Che–Castaldo J, Hodgson D, Scheuerlein A, Caswell H, Vaupel JW (2016) COMADRE: a global data base of animal demography. J Anim Ecol 85(2):371–384

Tacutu R, Craig T, Budovsky A, Wuttke D, Lehmann G, Taranukha D, Costa J, Fraifeld VE, de Magalhães JP (2013) Human ageing genomic resources: integrated databases and tools for the biology and genetics of ageing. Nucleic Acids Res 41(D1):D1027–D1033

Turkalo A, Klaus-Hügi C (1999) Group size and group composition of the bongo *Tragelaphus eurycerus* at a natural lick in the Dzanga National Park, Central African Republic. Mammalia 64(3):437–448

van de Kerk M, de Kroon H, Conde DA, Jongejans E (2013) Carnivora population dynamics are as slow and as fast as those of other mammals: implications for their conservation. PLoS ONE 8(8):1–9. doi:10.1371/journal.pone.0070354

Wich SA, Shumaker RW, Perkins L, de Vries H (2009) Captive and wild orangutan (*Pongo* sp.) survivorship: a comparison and the influence of management. Am J Primatol 71:680–686

Part V
More Topics

Chapter 18
Incomplete and Missing Data

Abstract Studbook data on wild animal species are generally of high quality. However, studbooks are not exempt from inaccuracies due to incomplete and missing historical records and/or difficulties in obtaining data, e.g. date of birth of young born in a den or parents in social groups. Information that is often missing includes sex, dates of birth/death and parentages. Data validation and guidelines for tracing information sources can reduce some inaccuracies, but not necessarily all. This chapter discusses the effect of incomplete and missing data on results of demographic and genetic analyses. Two–way contingency tables are applied to test whether observed sex–ratios in which unknown sex individuals are excluded differ from sex–ratios in which it is assumed that *all* unknown sex are either male or female. Incomplete dates in (SPARKS) studbooks refer to estimated by day, month or year or a quantified range around an expected date. Studbook guidelines recommend using the mid–point of date estimates: mid–month, mid–year or middle of breeding season. Effects of these estimates on seasonal analysis, median lifespan, age–class based mortality and fecundity and census analyses are discussed. The handling of unknown parentages in genetic analyses and the use of natural history data to trace potential parents are also covered.

18.1 Introduction

Wild animals in captivity are, in general, closely monitored on a daily basis. As a result, the data quality tends to be high. Moreover, animal record keeping systems, used widely in the zoo world since the 1980s (Flesness and Mace 1988), not only properly register data, but also provide tools to validate data. However, this does not mean that studbooks are exempt from inaccuracies. For example, a study involving ISIS data sets on 51 species, covering different taxonomic groups, showed that on average some 4 % of the dates of birth were not accurate (Kohler et al. 2006). However, it should be mentioned that this mainly referred to animals from unknown or wild origin.

This chapter uses the term *incomplete* to refer to partly available data, e.g. date of birth accurate by month, and *missing* to mean "unknown", e.g. unknown date or unknown sex. However, incomplete data can also be similar to missing data, e.g. a date accurate by month does not provide information for neonatal analyses.

F.P.G. Princée, *Exploring Studbooks for Wildlife Management and Conservation*, Topics in Biodiversity and Conservation 17, DOI 10.1007/978-3-319-50032-4_18

Zoo organisations provide guidelines for improving the quality of studbook data (see Thompson et al. 1997; Wilcken and Lees 2012). These guidelines are helpful in solving problems relating to incomplete and missing data. In spite of having access to guidelines and (electronic) validation tools, population managers still have to deal with studbooks where the data are only partly available. This may apply not only to older historical data, but also to (future) studbook data on wild, reintroduced or semi–wild populations, where monitoring of all individuals on a daily basis may not be feasible.

The main data that can be incomplete or missing and can have an impact on analysis results are sex, dates (birth, death, moves), parents and location of birth. This chapter focusses on handling these data, rather than solving their lack. Handling ranges from excluding unknown data, reducing the level of detail by including partial data, to making assumptions regarding unknown data (and including them). Each of these approaches has disadvantages. Excluding data reduces sample sizes and, thus, potentially also validity of analysis results. A lower level of detail may not allow the application of a required analysis, e.g. seasonality in the case that dates are estimated only by year. Assumptions regarding unknown data may be incorrect e.g. assuming an equal sex–ratio.

18.2 Unknown Sex

Various demographic analyses are usually applied to males and females separately, as species–specific differences between sexes often exist. This approach implies that the sex of individuals can be determined. This is obviously not the case in species with asexual reproduction. Sex–determination is also hindered in species without obvious external sexual dimorphism, whether during the juvenile stages or during the entire life–time.

Sex–ratio itself is obviously subject to effects of unknown sex. Excluding unknown sex reduces sample size, while the assumption that sex in the unknown group is distributed according to the observed sex–ratio may not be valid. One way to test the robustness of these data is by comparing the observed sex–ratio with the ratio based on the assumption that all unknowns are of the same sex.

Table 18.1 presents a two–way contingency table with males and females in a hypothetical sample of 100 births with ten individuals of unknown sex. The observed sex–ratio is equal. The Fisher's exact test (or Pearson's chi–square test) can

Table 18.1 Two–way contingency table with observed and assumed sex–ratio in sample of 100 births and 10 individuals of unknown sex. Fisher's exact test on sex–ratio: $p = 0.561$

	Males	Females	Total
Observed	45	45	90
Unknown = male	55	45	100

be applied to test whether the sex–ratio differs significantly when these unknowns are all assumed to be males. The null hypothesis that both samples have the same sex–ratio is not rejected ($p = 0.561$). In addition, the probability that ten individuals in which sex cannot be determined, are either all males or all females is low, in an equal sex–ratio model, i.e. $2 \times (\frac{1}{2})^{10} \approx 0.002$.

The age at which sex can be determined has a large impact on interpretation of mortality rates in neonate/juvenile age classes when computed per sex group. Excluding offspring which have been counted in a nest or den, but die and "disappear" before sex can be determined, results in underestimating mortality in the youngest age class. Consequently, only males and females which survived long enough to have their sex determined would be included in those analyses.

Fecundity rates are generally based on the number of births of the same sex as the parental group (see Chap. 7). Excluding offspring of unknown sex would result in underestimating fecundity rates in both male and female groups.

A common approach in studbook analysis is to assume equal sex–ratio and to partially count individuals of unknown sex as 0.5 in both males and females at risk of mortality, and in age distribution; and as a 0.5 male and female birth in fecundity (Faust et al. 2012b; Traylor–Holzer 2011). However, when binomial tests show that sex–ratios are unequal (see Chap. 5), the partial counts should be proportional. Binomial tests in polytocous species should only include litters with individuals of known sex.

Sex–ratio in the unknown sex group does not necessarily follow the observed sex–ratio. For example, sex–biased siblicide, as observed in spotted hyaenas (*Crocuta crocuta*) (Hofer and East 1997), can result in different sex–ratios among those that died and those that survived. Neonatal siblicide in this species has been observed in the den directly after birth (Frank et al. 1991), increasing the chance that the sex of the dead sibling remains unknown.

18.3 Incomplete Dates

Dates are required in analyses of natural history features such as seasonality and inter–birth interval, but above all in demographic analyses. This group of analyses is all about age, whether first age at reproduction, oldest observed age, age distribution or age–specific life tables. It is obvious that all of these analyses will be affected by inaccuracies in dates of birth and death.

Missing, i.e. unknown, dates do not provide any information for age–based analyses and are generally excluded from analyses. Fecundity tables are an exception, as parents of unknown age are treated as unknown parents (see next section). Unknown dates are likely to be found in historical zoo records, especially those prior to the introduction of computerised animal registration, and for individuals that have been obtained from the wild.

Incomplete calendar dates are those that are estimated by (nearest) day, month or year. Nearest day is generally not used (Wilcken and Lees 2012), but technically speaking occurs whenever a birth is not witnessed during the "out of zoo hours" period, i.e. between staff leaving and returning. Studbook guidelines recommend using the mid–point of date estimates: mid–month (day 15 or 16) and mid–year (1 July) or the mid–point of the breeding season for births (Thompson et al. 1997; Wilcken and Lees 2012).

To what extent incomplete data can be used in analyses depends on the type of analysis, level of resolution, and the species' biology. For example, using a monthly mid-point (e.g. 15 July) as the estimate for a birth known to have occurred in a particular month (July), means that the actual birthdate can at most be only 15/16 days from the mid-point. This is negligible when estimating longevity in species that live beyond five years – the error is $<1\,\%$. However, studies on perinatal and neonatal mortality require a resolution of one day, and so require accurate dates of birth and deaths.

Circular statistics as used in seasonality, can be applied to both individual dates and dates grouped by month (Zar 1984). The advantage of individual data is that the mean date can point to a potentially relevant date, e.g. mean birth around 21 June (summer solstice in Northern hemisphere). Although this resolution is lost when analysing grouped data, the advantage is that dates estimated by month can be included in the circular statistics, e.g. data from offspring that are not observed until they emerge from a den.

The probability that the true date of birth or death occurred on any day within the estimated interval is equal in the absence of seasonality. It is like random sampling from a pool of 30 or 365 day numbers. The mean of the sample will approximate the mid–point date (ignoring differences between months or even/uneven years).

However, the use of estimated dates comes at a cost by introducing additional stochastic uncertainty, i.e. the mean or median among true dates deviates from the mid–point date. Chapter 19 presents results of simulation experiments that illustrate these stochastic effects.

The effect of incomplete dates on age–class based life tables is illustrated with two individuals with estimated mid–year birth dates, which died on 1 May and 1 September of the next year, respectively. The ages of these individuals would be estimated as 304 and 427 days, respectively. Since true birth dates of these individuals can be anywhere between 1 January and 31 December in a birth–flow model, their true ages at death could range from 121 – 486 and 244 – 609, respectively.

Age class values are truncated to the lowest. The age ranges are not evenly distributed around the 365 days. This means that the probability of entering class 1 is not equal between (random) dates of birth. Figure 18.1 shows ages at death of both individuals for different dates of birth. The bottom line represents death on 1 May, the top line death on 1 September. The age is presented in years in order to visualise the class number i.e. age less than 1 year (dotted horizontal line) is class 0.

The area left and above the intercept between age and the 1.0 year line represents the probability of having died in age class 1; right and below refers to class 0. Unequal areas imply unequal probabilities. These probabilities depend on date of death, and are equal at the mid–point.

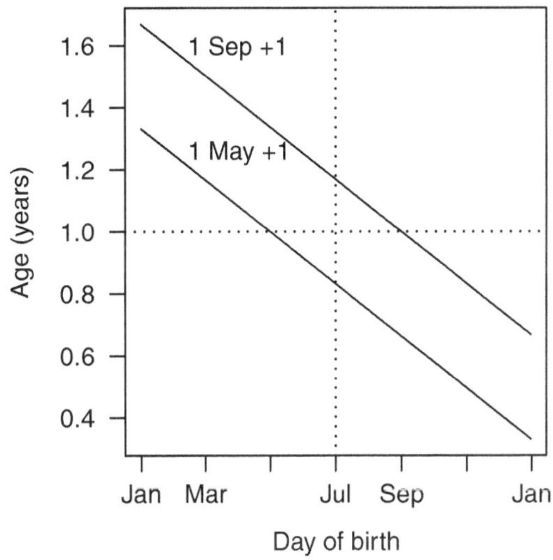

Fig. 18.1 Ages at 1 May and 1 September in the next year for each date of birth. The *vertical dotted line* refers to mid–year birth date (estimated)

The above example was based on estimated dates of birth. However, the same effects also apply to estimated dates of death or, in fecundity, to estimated dates of birth of offspring. The total effect of estimated dates on life tables requires further studies on real populations.

Census counts do not depend on exact dates as long as individuals were born/arrived or died/left the population within the census interval. This means that mid–year date estimates can be included in "end–of–year" census, which is often applied to birth–flow populations. However, pre– and post–breeding census of birth–pulse populations are affected by mid–year date estimates. Since breeding seasons do not necessarily start in January, the census interval is not the same as a Gregorian calendar year. Therefore, mid–year estimates are not the same as mid–points of the census interval. The effect is similar to that discussed for age–class based life tables.

18.3.1 Unknown Parents

Age at first breeding, inter–birth interval and reproductive lifespan analyses (see Chap. 3) only include individuals listed in the studbook that have reproduced. Fecundity tables, however, can distribute offspring of unknown parents according to the observed distribution of fecundity rates (m_x) (Faust et al. 2012a).

Missing parentages mostly impact genetic analyses: inbreeding coefficients, founder representation, kinship values and estimates of genetic loss all depend on information on parentages in the studbook pedigree. Three categories of unknown parents can be recognised:

1. Parents of wild–born individuals.
2. Parents of individuals born at unknown locations.
3. Unknown parents of captive–born individuals of known location.

The convention is to assume that wild–born individuals are unrelated to each other i.e. they are randomly selected from an infinitely large population. That assumption will not easily hold for small endangered populations, where many individuals may already be related. Furthermore, imports of wild individuals of species with clutches or litters can involve siblings – information which was not always provided in the past. Molecular genetics are required for information on relatedness between wild–born individuals, especially when they are potentially siblings.

Individuals born at unknown locations have almost by definition unknown parents. These are the "real" unknown parents, as they might be wild parents or captive–born parents. These individuals can be considered as founders if they entered the studbook population in the period before regular breeding started in captivity. Otherwise, molecular genetics are required to assess their relatedness to other individuals.

Studbooks indicate whether individuals with wild or unknown parents are known to be (half–)siblings, by adding a unique number to the *WILD* or *UNK* IDs of the parent(s). The unique coding of parents of (potential) founders can be used in studbook software, such as PMx (Ballou et al. 2010), to provide better estimates of relatedness between individuals (and thus of inbreeding coefficients, founder representation, gene diversity, etc.).

Individuals which are captive–born but have unknown parents affect the results of genetic analyses, especially when they have reproduced. The earlier GENES program included the option of omitting individuals with unknown parents or of considering them as founders (Lacy 1993). However, treating these individuals as founders is not acceptable if it is clear that they are captive–born and their potential parents are known e.g. males in multi–male groups or females in herds. Multiple possible parents are generally coded as *MULT* with an optional unique number to distinguish between social groups or herds.

Multiple parents were interpreted as single unknown parents in genetic analyses. However, this has changed with the release of the PMx program which can include multiple possible parents in the kinship calculations (Ballou et al. 2011; Lacy 2012). The SPARKS program v1.6 offers the option of specifying multiple potential parents with (optional) their probabilities of being "the" parent (ISIS 2012).

A recent development is genetic management based on pedigree analysis of group–living organisms using the features of PMx to handle unknown and multiple parents (Jiménez-Mena et al. 2016). This type of management can handle irregular pedigrees and has, therefore, great advantage over *Maximal Avoidance of Inbreeding (MAI)* schemes, which require structured patterns to exchange individuals between social groups (Princée 1995, 1998).

Natural history data, such as reproductive lifespan, gestation length, inter–birth interval, in combination with reproductive history of individuals, can be used to determine which individuals are possible parents. This approach was applied to trace potential parents of individuals in a herd of Przewalski's horses (*Equus przewalskii*) (Princée 1998).

Tracing potential parents can be tedious in larger studbooks with a high proportion of older historical records or in studbooks of species living in social groups. The PML software traces potential parents at location of birth using results from natural history analyses (Princée 2014). However, natural history data are species–specific and can not be fully implemented in software. For example, females may give birth outside the breeding season after stillbirth or neonatal death in the zoo environment. This means that human judgement is required to make reasonable choices.

Molecular genetic methods such as DNA fingerprinting and mitochondrial DNA analyses are useful in detecting errors in registered parentages and in solving unknown parents in pedigrees (e.g. Bowling et al. 2003; Morin and Ryder 1991). However, these techniques are not a guarantee that all problems are solved. Genetic material of founders and individuals in previous generations are not always available, particularly for older studbook populations. For example, a genetic study involving 51 loci of blood type and DNA markers could only partly solve the unknown parentages in the Przewalski's horse herd as described above (Bowling et al. 2003). This means that even in "complete" studbooks one may have to accept a certain number of unknown parentages.

References

Ballou JD, Lees C, Faust LJ, Long S, Lynch C, Bingaman Lackey L, Foose TJ (2010) Demographic and genetic management of captive populations. In: Kleiman DG, Thompson KV, Kirk Baer C (eds) Wild mammals in captivity: principles and techniques for zoo management, 2nd edn. The University of Chicago Press, Chicago, pp 219–252

Ballou JD, Lacy RC, Pollak JP (2011) PMx: software for demographic and genetic analysis and management of pedigreed populations (version 1.0). Technical report, Chicago Zoological Society, Brookfield

Bowling AT, Zimmerman W, Ryder O, Penado C, Peto S, Chemnick L, Yasinetskaya N, Zharkikh T (2003) Genetic variation in Przewalski's horses, with special focus on the last wild caught mare, 231 Orlitza III. Cytogenet Genome Res 101:226–234

Faust LJ, Bergstrom YM, Thompson SD, Bier L (2012a) PopLink version 2.4. Lincoln Park Zoo, Chicago

Faust LJ, Bier L, Schowe K, GazLay T (2012b) PopLink version 2.4: user's manual. Lincoln Park Zoo, Chicago

Flesness NR, Mace GM (1988) Population databases and zoological conservation. Int Zoo Yearb 27:42–49

Frank LG, Glickman SE, Licht P (1991) Fatal sibling aggression, precocial development, and androgens in neonatal spotted hyenas. Science 252(5006):702–704

Hofer H, East ML (1997) Skewed offspring sex ratios and sex composition of twin litters in Serengeti spotted hyaenas (*Crocuta crocuta*) are a consequence of siblicide. Appl Anim Behav Sci 51(3-4):307–316

ISIS (2012) SPARKS 1.66: Single Population Animal Record Keeping System. International Species Information System, Eagan

Jiménez-Mena B, Schad K, Hanna N, Lacy RC (2016) Pedigree analysis for the genetic management of group-living species. Ecol Evol 6(10):3067–3078
Kohler IV, Preston SH, Lackey LB (2006) Comparative mortality levels among selected species of captive animals. Demogr Res 15:413–434
Lacy RC (1993) Genes: a computer program for the analysis of pedigrees and genetic management of populations. Software, Chicago Zoological Society, Brookfield
Lacy RC (2012) Extending pedigree analysis for uncertain parentage and diverse breeding systems. J Hered 103(2):197–205
Morin PA, Ryder OA (1991) Founder contribution and pedigree inference in a captive breeding colony of lion–tailed macaques, using mitchondonrial DNA and DNA fingerprint analyses. Zoo Biol 10:341–352
Princée FPG (1995) Overcoming the constraints of social structure and incomplete pedigree data through low-intensity genetic management. In: Ballou JD, Gilpin M, Foose TJ (eds) Population management for survival and recovery: analytical methods and strategies in small population conservation. Methods and cases in conservation science series. Columbia University Press, New York, pp 124–154
Princée FPG (1998) Genetic management of small animal populations in zoos and wildlife reserves: the use of genome models in the estimation of genetic variation and the effects of social structures. Ph.D. thesis, University of Groningen, Groningen
Princée FPG (2014) Population Management Library (PML) Software and manual version 1.0. East Rudham. http://www.princee.com
Thompson SD, Earnhardt JM, Turner-Erfort G (1997) Guidelines for data entry and maintenance of North American regional studbooks. Lincoln Park Zoo, Chicago
Traylor–Holzer K (ed) (2011) PMx users manual, version 1.0. IUCN SSC Conservation Breeding Specialist Group, Apple Valley
Wilcken J, Lees C (eds) (2012) Managing zoo populations: compiling and analysing studbook data, 3rd edn. Australasian Regional Association of Zoological Parks and Aquaria. http://training.isis.org/CommunityLibrary/ManagingZooPopulations-2012edition.pdf
Zar JH (1984) Biological statistics, 2nd edn. Prentice-Hall, Englewood Cliffs

Chapter 19
Statistical Topics

Abstract Background information and experiments involving some mathematical/statistical topics that were only briefly touched on in previous book chapters are presented. The first topic covers the use of circular statistics in seasonality analysis e.g. detecting the peak in births. The estimated mean day or month based on the numbering sequence of the human calendar can be incorrect when seasons continue into the next year. This requires adjusting the sequence of months/days so that the start of season has the lowest number. Circular statistics do not require a true zero point and are therefore suitable for seasonality analysis. The breeding season of Nepalese red pandas in the southern hemisphere is used as an example to compare circular statistics with calendar–based analysis. The topic of pseudoreplication in life history analysis has not received much attention in studbook management. Experiments with bootstrapping show the impact of litter, maternal and spatial (location at birth) effects on variance in 30–day mortality in snow leopards. Estimated dates of birth and death are entered as mid–points of the interval (month, year). Simulation experiments are used to study the effects of different proportions of estimated ages on the median lifespan. These effects result in wider variances, thus higher uncertainty, in median lifespan. The last topic presents a method to transform age–class specific mortality rates from studbook data into a constant (adult) mortality rate, which is used in the VORTEX simulation model. The chapter ends with some remarks on using statistics in studbook analyses.

19.1 Introduction

Exploring in the context of this book not only refers to revealing biological information which is embedded in studbook data, but to statistical methods and techniques to validate this information as well. The open source program (and language) *R* (R Core Team 2015) provides a wealth of statistical tests and graphic tools to explore data. Access to nearly 9,000 contributed packages (on 1 August 2016) can be overwhelming, and may make it hard to decide which tests are appropriate.

Statistical issues with data that do not follow a "known" distribution (e.g. normal or Poisson) or which are not independent, have been discussed in the

F.P.G. Princée, *Exploring Studbooks for Wildlife Management and Conservation*,
Topics in Biodiversity and Conservation 17, DOI 10.1007/978-3-319-50032-4_19

relevant chapters. This chapter will provide background information and experiments involving some mathematical/statistical topics that were only briefly touched on in previous book chapters.

Circular statistics are preferred in analysis of seasonality in births and deaths. The potential pitfall of applying the well known Pearson's chi–square test on seasonal data is discussed in Sect. 19.2.

Studbooks are collections of data from different locations (zoos) and covering several years, or even decades. Therefore it is unsurprising that studbook data do not necessarily represent a random sample. Section 19.3 illustrates the effects of dependent data, which is a type of pseudoreplication (Hurlbert 1984), on neonatal mortality using bootstrap methods.

A simulation study to show effects of assumptions regarding estimated (incomplete) dates of birth or death on life tables is presented in Sect. 19.4. The last topic in this chapter shows how to estimate a constant mortality rate from multiple age classes as used in the VORTEX software (Lacy and Pollak 2014) (see Sect. 19.5).

19.2 Circular Statistics

Seasonality data can be graphically represented in a histogram. Figure 19.1a presents a histogram of 119 litter dates grouped by month in Nepalese red pandas in southern hemisphere zoos during the period 1977–2012. This figure shows that the season is November to January, with December being the peak month.

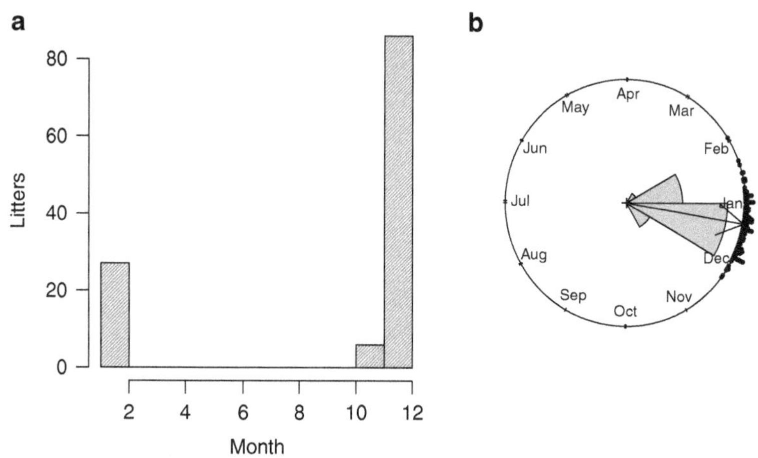

Fig. 19.1 (**a**) Histogram of litters grouped per month in Nepalese red pandas in the southern hemisphere from 1977 to 2012. (**b**) Circular graphics of the same data. Stacked points refer to litters per day of year of the same data; the area sectors in the rose diagram represent relative frequencies of litters per month; and the *arrow* indicates the mean litter date (21 December)

Calculation of mean season from day of year (1–365) or month numbers (1–12) can yield erroneous results when the season starts in one year and end in the next. In this example, mean day would be 273 (29 September) and the mean month is 9.46 (mid–September). However, there are no observations in September (Fig. 19.1a). A solution to this problem is to re–code day and month numbers to the start of the season e.g. 1 November is the first day of the "year". A better approach is to use statistics that have been developed to handle data with no true zero points.

Circular statistics can be applied when the interval scale of measurement does not have a true zero point, and where the designation of high and low values are arbitrary (Zar 1984). Seasonality, especially in the case when events continue after the start of a new (astronomic or human) calendar year, is a good example of the application of this method.

Dates, whether numbered in day, week or month of the year, are converted to an angular direction (a) in degrees or radians, on a circular scale of 360° (or 2π), which is divided into equal time intervals (Zar 1984). For example, data grouped per month[1] will be $k = 12$ when dates are converted to intervals, in this case of 30°, corresponding to the month of the year.

Circular data are often presented in circular (data) plots and rose diagrams (Pewsey et al. 2013). Figure 19.1b presents the red panda data in circular data plots with stacked bars as frequency counts of litters per day. The area sectors of the rose diagram represent relative frequencies of monthly data. The mean angle $\bar{a}$ and the length of the mean (resultant) vector r are calculated from these data. The mean day or month can be obtained by converting $\bar{a}$ back to the appropriate time units.

The significance of $\bar{a}$ can be tested with the *Rayleigh test* (see Zar 1984). The null hypothesis H_0 of this test is: "The sampled data are distributed uniformly around the circle". When this hypothesis is rejected, the data are *not* uniformly distributed, and the mean angle $\bar{a}$ refers to the mean day or month. Circular statistics on the red panda data result in 21 December being the mean day of birth ($\bar{a} = 350.1°$). The arrow in Fig. 19.1b represents the mean date.

The Rayleigh test rejects the null hypothesis ($r = 0.98$, $p < 0.0001$, $n = 119$). Alternative tests are the *Kuiper's one sample test* and the *Watson one sample test* (Zar 1984). The *Watson two–sample test* is used to test whether two samples are from the same population (Zar 1984).

The Rayleigh, Kuiper's and Watson tests for uniformity do not reveal multiple seasons. For example, the length of the vector (r) will be small in the case of seasons with a 6 month interval, so graphic presentations of the data are important for visually detecting multiple seasons.

The PMx program applies the chi–square to test whether births and deaths, grouped per month, are distributed uniformly over the year (Lacy et al. 2012). Since

[1]The conversion of months to 12 equal intervals assumes that the number of days is 30 days in each month.

this method does not estimate the mean month, re–coding of month numbers is not required in the case the season continues in the next year.

19.3 Pseudoreplication

> Pseudoreplication is defined as the use of inferential statistics to test for treatment effects with data from experiments where either treatments are not replicated (though samples may be) or replicates are not statistically independent (Hurlbert 1984).

The issue of pseudoreplication in zoo research was not addressed until a literature study was carried out a decade ago (Kuhar 2006). The problem of small sample sizes, e.g. a single pair or group per zoological institution, "forces" researchers to pool sample data from different institutions. However, there is a "danger" attached to pooling, as individuals from the same zoological institution share the same fate by sharing the same environment (e.g. enclosure, local climate) or treatment (e.g. diet). This means that the pooled data are not statistically independent.

The total (pooled) sample size in this type of pseudoreplication is smaller than the sum of the zoo samples, from a statistical point of view. This results in underestimation of the variance (which should be estimated using the pooled sample size), which can lead to finding statistically significant differences in a study when no differences actually exist (i.e. Type I error).

International and regional studbooks of endangered species in zoos are, de facto, pooled data and therefore subject to underestimation of variances. Statistics to model demographics and genetics were not included in Kuhar's (2006) study. This does not mean that these models are not prone to pseudoreplication.

Pseudoreplication in mortality occurs when litter mates, siblings (from multiple litters) or individuals born at the same location, share the same fate, i.e. the chance of dying is higher or lower than would be expected from the mortality rate in the studbook population. Examples include litter mates that die due to a disease, a female who neglects her offspring (multiple litters), or a sub–optimal enclosure.

Bootstrap experiments on 30–day mortality (q_{30}) in snow leopards are used to illustrate the effect of pseudoreplication. This resampling technique was introduced in Chap. 8 to estimate variance in age–specific life tables. The method implemented in the PML software samples individuals at risk of dying (Princée 2014). The bootstrap experiments in this section also sample litters, mothers (dam) and locations at birth. The individuals in these sample units form the pooled mortality data. Since bootstrap sampling is with replacement, units can occur more than once. If mortality within units is random, one would not expect differences in variance between bootstrap of individuals (also observed data) and litter, dam or location.

Table 19.1 presents the q_{30}, variance, standard error (SE), confidence interval (percentiles) and coefficient of variation (CV) in observed data and bootstrap experiments. The observed q_{30} and mean q_{30} values in bootstrap experiments

Table 19.1 Observed data and bootstrap experiments on 30 day mortality in snow leopards. Data are right censored. Number of runs is 999

Group (N)	q_{30}	Variance	SE	Confidence interval[a] Lower	Upper	CV (%)
Observed (1,451)[b]	0.286	0.0001	0.0119	0.263	0.309	4.15
Individuals (1,451)	0.285	0.0001	0.0116	0.263	0.308	4.05
Litter (671)	0.286	0.0002	0.0148	0.256	0.313	5.19
Dam (268)	0.285	0.0004	0.0191	0.247	0.322	6.70
Birthsite (155)	0.286	0.0004	0.0199	0.248	0.325	6.94

N number of "elements" in group, q_{30} 30 day mortality (mean in bootstrap), *SE* standard error of mean, *CV* coefficient of variation

[a]Confidence intervals are expected 95 % values in a binomial distribution for observed data and 2.5–97.5 percentiles in bootstrap experiments

[b]Variance, standard error and coefficient of variation are expected values of the binomial distribution

do not differ. However, the variance, and consequently standard error (SE) and coefficient of variation increases with larger but fewer "units", i.e. individual, litter, dam and location. The consequence of larger variances, is wider confidence intervals. The coefficients of variation are a good indicator of the effects of pseudoreplication.

Statistical software programs that have been developed since Stuart Hurlbert published his paper on pseudoreplication in 1984 include various techniques to handle dependent data. This particularly refers to generalised linear regression models (GLMs) which can handle dependent data as random effects. The use of marginal (cluster) and frailty (random) models in survival analysis is an example (Wienke 2010). See also Chap. 8.

19.4 Estimated Age

Monte Carlo simulation experiments are useful to assess effects of dates of birth and/or death that are estimated as "mid–month" or "mid–year". This section describes an experiment to assess the effect of estimated ages on median lifespan. The experiment involves five populations with different proportions of age at death that are based on estimated dates of birth.

The base population contains accurate ages at death of snow leopards in the international studbook that survived the first month ($n = 877$). Four populations ($A - D$) are based on the same data, but respectively 5, 10, 25 and 50 % of the death–ages above 30 days are assumed to be estimates based on mid–month dates of birth. The fifth population (E) is a *single* random sample without replacement of 100 records from the base population in which 50 % of the ages above 30 days are assumed to be an estimate.

Table 19.2 Simulation experiment on effects of estimated dates on median lifespan of snow leopards that survived the first 30 days ($N = 877$). Population E is a random sample of 100 of the original population. Number of runs is 9,999

ID	%	N_{est}	Median lifespan (in days)			s^2
			Observed data	Estimates excluded	Simulation[a]	
A	5	43	3,085	3,131	3,085	0.43
B	10	86	3,085	3,123	3,086	50.44
C	25	216	3,085	2,966	3,086	54.03
D	50	433	3,085	3,094	3,085	2.47
E[b]	50	50	3,281	3,206	3,281	2.95

ID population ID, *%* percentage of records with estimated age at death > 30 days, N_{est} number of records with estimated ages, *Observed data* median lifespan as observed in the population, *Estimates excluded* median lifespan in the population when estimated ages are excluded from the calculation, *Simulation* median of simulation values for median lifespan, s^2 variance in median lifespan among 9,999 runs

[a]Median of medians over 9,999 runs

[b]Population is random sample of 100 from group that died after 30 days

Random sampling without replacement was applied to draw the group of individuals with assumed estimated ages for each population ($A - E$). As this sampling process was carried out once, the distribution of estimated dates within each population is unknown.

The real age calculated from a mid–month date of birth can range, with equal chance, from $age - 15$ *to* $age + 15$ days (depending on days per month). Monte Carlo methods were used to determine the "real" age by drawing the number of days to adjust from the range $[-15, 15]$ for each estimated age. The median lifespan was estimated after these new ages were drawn. This process was repeated during each of the 9,999 runs. The median and the variance for lifespan were calculated at the end of the simulation.

Table 19.2 presents the results of these simulation experiments. The median lifespan in the original (accurate) data is 3,085 days. The medians of simulation values for median lifespan do not differ from the observed value(s). This result is to be expected given the equal chance that the actual age is less or more than the estimated age.

Excluding estimated ages from calculations results in median values that are different from including them. These differences are not only due to the proportion of estimated ages, but also to the distribution of estimates over different age groups. The populations are single random samples for each "proportion" group. This explains why excluding estimates in population C results in a lower median lifespan than in D. The important message, however, is that excluding estimated dates by month has more effect on the median lifespan than including them.

Variance in median lifespan is expected to be higher in populations with large proportions of estimated dates and/or smaller population sizes. However, the distribution pattern of estimated dates also contributes to the variance in median lifespan, as illustrated in populations B, C and D. More detailed simulation experiments are required to assess the contribution of both variance components.

19.5 Constant Adult Mortality

The simulation program VORTEX that was used in the lowland bongo PVA (see Chap. 17) uses a constant, age–bound, mortality class for adult reproductive individuals (q_{adult}). However, studbook software generally calculates mortality rates for each age class separately. A method to estimate constant mortality from individual age classes is presented in this section.

A constant adult mortality results in geometric change (decrease) in numbers of individuals at risk of dying per age class. The following equation applies:

$$N_{x+n} = N_x(1 - q_{adult})^n \tag{19.1}$$

where n is the number of classes, N_x and N_{x+n} are numbers of individuals at risk for classes x and $x+n$, respectively, and q_{adult} is a constant mortality rate. Note that this equation can be simplified by using the constant survival rate (p_{adult}).

Values for individuals at risk (N_x) can be directly obtained from life tables when all individuals belong to the same cohort (i.e. born at the same time). Equation 19.1 is not valid when life tables are based on staggered data entry (see Chap. 7). The remedy to this problem is to estimate the constant survival rate p_{adult} from survivorship (l_x):

$$l_{x+n} = l_x(p_{adult})^n \tag{19.2}$$

where n is the number of classes, l_x and l_{x+n} are survivorship for age classes x and $x+n$, respectively. Solving Eq. 19.2 requires a log transformation:

$$\log l_{x+n} = \log l_x + n(\log p_{adult}) \tag{19.3a}$$

$$\log p_{adult} = (\log l_{x+n} - \log l_x)/n \tag{19.3b}$$

$$p_{adult} = antilog\left[(\log l_{x+n} - \log l_x)/n\right] \tag{19.3c}$$

For example, survivorship for l_5 and l_{15} are 0.473 and 0.050. The number of classes n is 10. The natural logarithm ($\log_e$ or ln) is used in the log transformation in this example. The constant mortality rate (q_{adult}) for classes 5–15 is calculated following Eq. 19.3:

1. $\ln p_{adult} = (\ln 0.050 - \ln 0.473)/10 = -0.223$
2. $p_{adult} = e^{-0.223} = 0.799$
3. $q_{adult} = 1 - 0.799 = 0.201$.

Thus, the constant mortality rate in (adult) age classes 5–15 is 0.20 (20 %).

19.6 Remarks

The statistical analyses in this book vary from descriptive statistics such as Tukey's five number summary (Tukey 1977) to advanced (inferential) statistics such as Cox's regression model (Cox 1972) and the animal model (see Hadfield 2010; Vazquez et al. 2010).

Exploring studbook data with descriptive statistics, particularly graphic representations, is recommended. The power of these statistics is to provide understanding of characteristics and/or trends in studbook data, free from assumptions regarding the distribution of the data. However, descriptive statistics do not replace inferential statistics as one needs to know whether the observed trend is a "true" trend or can be explained by random chance alone.

Conditions and interpretation of results from advanced inferential statistics can be a challenge, e.g. Cox's regression model (Cox 1972). Therefore, it is recommended to first apply less advanced models in order to gain understanding of realistic values.

Last but not least, understanding of the (biological) features and environmental factors that can affect studbook data is essential for deciding which analyses are appropriate to apply.

References

Cox DR (1972) Regression models and life–tables. J R Stat Soc Ser B (Methodol) 34(2):187–220

Hadfield JD (2010) MCMC methods for multi-response generalized linear mixed models: the MCMCglmm R package. J Stat Softw 33(2):1–22

Hurlbert SH (1984) Pseudoreplication and the design of ecological field experiments. Ecol Monogr 54(2):187–211

Kuhar CW (2006) In the deep end: pooling data and other statistical challenges of zoo and aquarium research. Zoo Biol 25(4):339–352

Lacy RC, Ballou JD, Pollak JP (2012) PMx: software package for demographic and genetic analysis and management of pedigreed populations. Methods Ecol Evol 3(2):433–437

Lacy RC, Pollak JP (2014) VORTEX: a stochastic simulation of the extinction process, version 10.0. Technical report, Chicago Zoological Society, Brookfield

Pewsey A, Neuhäuser M, Ruxton GD (2013) Circular statistics in R. Oxford University Press, Oxford

Princée FPG (2014) Population Management Library (PML) software and manual version 1.0. East Rudham. http://www.princee.com

R Core Team (2015) R: a language and environment for statistical computing. R foundation for statistical computing, Vienna. https://www.R-project.org/

Tukey JW (1977) Exploratory data analysis. Behavioural science: quantitative methods. Addison-Wesley, Reading

Vazquez AI, Bates DM, Rosa GJM, Gianola D, Weigel KA (2010) Technical note: an R package for fitting generalized linear mixed models in animal breeding. J Anim Sci 88(2):497–504

Wienke A (2010) Frailty models in survival analysis. Biostatistics series. Chapman and Hall/CRC, Boca Raton

Zar JH (1984) Biological statistics, 2nd edn. Prentice-Hall, Englewood Cliffs

Appendix A
Software

This appendix contains overviews of commands and functions that are available in the *Population Management Library* and the *studbookR* package. Analyses in this book have been carried out with this software.

A.1 Population Management Library

The Population Management Library consist of modules for demographic and genetic analyses of studbook populations, written in C++11.[1] The library imports SPARKS studbooks that have been converted to SQLite3 databases.

This section contains the basic commands (and syntax) of the PML script language that were used to carry out the analyses in this book.

A.1.1 Commands

Single line comments in PML scripts are preceded by a hash (#) or double slashes (//).

```
set verbose on                        // Display messages

# Load project configuration (which includes name and format
# of studbook). This procedure can be skipped if the project
# name is passed to as command-line parameter
```

[1]ISO C++ standard ratified in 2011.

F.P.G. Princée, *Exploring Studbooks for Wildlife Management and Conservation*, Topics in Biodiversity and Conservation 17, DOI 10.1007/978-3-319-50032-4

```
load project "studbook"             // Load project data
load studbook                       // Load studbook data

#  Date settings

set date format to iso8601          // Date format: YYYY-MM-DD
set begin date to 1977-01-01        // Begin of time window
set end date to 2012-12-31          // End of time window
set monitor date to 2012-06-30      // Last date of monitoring
set census day to 30                // Census day: 30
set census month to 6               // Census month: June
set census interval to annual       // Annual census
set date accuracy all to day        // All dates:
                                    // accuracy by day

# Compute generation and inbreeding

compute generation                  // Captive--born generation
compute inbreeding                  // Inbreeding coefficients

set neonatal age to 30.0 days       // End of neonatal period
include neonates                    // Include neonates

export pedigree                     // Export pedigree for use
                                    // in R

# Set location view

clear location view                 // Clear location list
init location view                  // Initialise new list

# Geographic regions in northern hemisphere
set population to 'Europe'
set population to 'North America'
set population to 'Asia'

# Natural history data analyses

set sex to all                      // Or: male, female, unknown
compute longevity                   // Longevity and
                                    // median lifespan
compute birth season stillbirth     // Seasonality in births
                                    // includes stillbirths
compute litter season               // Seasonality in litters
compute death season                // Seasonality in death
```

```
set sex to male                     // Or: female
compute first breeding              // Age at first breeding
compute reproductive life           // Reproductive lifespan

# Crude demographic analyses

set sex to all                      // Or: male, female, unknown
set date accuracy all to month      // Allow estimates by month

compute census                      // Living at census date
compute births                      // Births in census interval
compute deaths                      // Deaths
compute neonatal deaths             // Neonatal deaths
compute immigration all             // Immigration (import)
compute immigration wild            // Import from wild

# Demographics

set date accuracy all to day
set class width to 1.0 years        // 365.25 days
set max classes to 20               // Pool >= 20 classes

# Data censoring and prorating settings
set bootstrap off                   // default: on
set resamples 1999                  // default: 999
set left truncated off              // default: on
set right censored off              // default: on
set prorating on                    // default: off
set handle unknown to ignore        // default: equal
set handle parent to ignore         // default: ratio

set sex to male                     // Alter for female, unknown
compute age distribution            // Age distribution at
                                    // monitoring date
compute mortality                   // Age-specific mortality

set sex to male                     // or: female
compute fecundity                   // Age-specific fecundity
compute lifetable                   // Full lifetable
                                    // e.g. Vx, Cx

# Data export for external Kaplan-Meier or
# Cox proportional hazard regression
```

```
set sex to male                      // or: female
export survival alive                // Survival data
                                     // excluding stillbirths


set sex to female
set rearing to parent                // Include parent-reared
exclude neonates                     // Exclude neonates
set origin to all                    // Origin are all groups

compute interbirth interval          // Inter-birth interval

# reset
set origin to captive                // Origin is captive-born
set rearing to all                   // Rearing is all types
include neonates                     // Include neonates

compute littersize                   // Litter size

set sex to all
compute founder group                // Founder representation
compute mean kinship                 // Mean kinship

gene drop


# Lethal equivalents

set sex to all                       // or: male, female,
                                     // unknown offspring

compute viability 180.0 noninbred // Survival to 180 days
                                     // both parents noninbred
```

A.2 studbookR

The R package *studbookR* contains functions for statistical analyses and graphic display of demographic and genetic data that have been exported by the Population Management Library (PML). The following section contains an overview of the main functions, with their default settings, that have been used to analyse and produce graphic figures for this book.

A.2.1 Functions

Short information on functions is provided as comments (behind #).

```
library(studbookR)                  # Load package studbookR
studbook()                          # Select studbook project

# Natural history analyses

# Functions in this group analyse the distribution of raw data

studbook.lifespan()                 # Age at death
studbook.longevity()                # Longevity
studbook.firstBreeding()            # Age at first breeding
studbook.breeding()                 # Offspring per age group
studbook.birthInterval()            # Inter-birth interval
studbook.litterSize()               # Litter size

# Circular statistics is applied in seasonality analyses.

studbook.season('birth')            # Seasonality in births
studbook.season('litter')           # Seasonality in litters
studbook.season('death')            # Seasonality in deaths

# Crude demographic analyses
# Graphic representation of census and census-style analyses;
# Calculation of growth, birth, death and migration rates

studbook.census()                   # Annual census
studbook.births()                   # Births between census dates
studbook.deaths()                   # Deaths between census dates
studbook.neonatalDeaths()           # Neonatal deaths between
                                    # census dates
studbook.immigration()              # Annual immigration (import)
studbook.emigration()               # Annual emigration (export)
studbook.ecology()                  # Fit logistic/exponential
                                    # model

# Life tables

studbook.age()                      # Age distribution last date
studbook.mortality()                # Age specific mortality
studbook.fecundity()                # Age specific fecundity
studbook.lifetableFull()            # Full life tables
                                    # (e.g. Vx, R0, r)
```

```
# Kaplan-Meier estimator and Cox proportional hazard regression
studbook.survival(method='kap')    # Kaplan--Meier survival
studbook.survival(method='cox')    # Cox regression

# Genetics

studbook.founders()                # Plot founder representation
studbook.inbreeding()              # Various inbreeding tests
studbook.genedrop()                # Distribution gene drop data


# Quantitative genetics

phenotype()                        # Select life history trait
phenotype.repeatability()          # Repeatability in trait

# Parent-offspring regression
phenotype.regression ()            # mid-parent  offspring
phenotype.regression (parent='dam',offspring='female')

# Animal model
pedigree.read()                    # Read pedigree file

# pedigreemm REML method

# default model: animal as random effect (1|animal)
animalmodel.reml()

# additional sex as fixed effect, litter as random effect
animalmodel.reml( "(1|animal) + SEX + (1|LITTER_ID)" )

# MCMCglmm Markov chain Monte Carlo method

# default model: animal as random effect
# improper prior
# default values for iterations, burnin, thinning
animalmodel.mcmc()

animalmodel.setFixed("SEX")        # sex as fixed effect
animalmodel.setPrior("observed") # observed variance
animalmodel.setRandom("animal + LOCATION") # animal and location
```

```
# iterations, thinning and burnin' period
animalmodel.setRuns(25000,25,5000)
animalmodel.mcmc()               # new model/settings
```

A.3 Availability

The Population Management Library and the studbookR package can be downloaded under an open source license from the software download area at http://www.princee.com.

Index

F.P.G. Princée, *Exploring Studbooks for Wildlife Management and Conservation*, Topics in Biodiversity and Conservation 17, DOI 10.1007/978-3-319-50032-4

Kolibri 360 GmbH
Ferdinand-Jühlke-Straße 7
99095 Erfurt, Deutschland
produktsicherheit@kolibri360.de